How The World Really Works

Table of Contents

Introduction 5

Chapter 1: The Universe Unveiled 22

    1.1   From the Cosmos to Our Cosmic Address 22

    1.2: The Birth and Fate of Stars 38

    1.3: Galaxies, Black Holes, and Dark Matter 72

Chapter 2: The Solar System and Planetary Mysteries 129

    2.1 A Journey through Our Solar Neighborhood: 129

    2.2 The Wonders of Our Planetary Neighbors: 148

    2.3 Mysteries of the Solar System: Unveiling the Cosmic Enigmas 178

Chapter 3: The Earth and Its Systems 205

    3.1: Geology and Plate Tectonics 205

    3.2: Earth's Climate and the Water Cycle .. 226

    3.3: The Dynamic Biosphere 243

Chapter 4: Evolution & Biodiversity 262

    4.1: The Theory of Evolution 262

    4.2: The Tree of Life: Biodiversity and Ecosystems 286

4.3: Genetics and Heredity: The Code of Life 306

Chapter 5: Genetics and Heredity..................323

   5.1 DNA and Genes 323

   5.2 Genetic Engineering............................329

Chapter 6: The Story of Life 346

   6.1: The Origin of Life..................346

   6.2: Early Life Forms..................368

   6.3: Evolutionary Milestones 382

Conclusion..397

Introduction

Exploring the Mysteries of Our World

Welcome to a journey of profound discovery, a voyage into the heart of our world—an exploration that transcends the boundaries of our everyday perceptions to uncover the intricate mechanisms, profound mysteries, and breathtaking wonders that define our existence. In "How the World Really Works," we embark on an odyssey that spans the cosmos, unravels the secrets of life, delves into the realms of quantum physics, and plumbs the depths of the human experience. This book is an homage to the insatiable curiosity that drives us to ask questions, to seek answers, and to find solace in understanding the world that surrounds us.

At first glance, the world appears familiar, its landscapes and skies, its oceans and forests, its bustling cities and tranquil villages. But beneath this veneer of familiarity lies a realm of extraordinary complexity and astonishing beauty. The purpose of this book is to peel back the layers of familiarity, to reveal the astonishing intricacies that underpin the world we inhabit. It's a journey that takes us from the grandeur of the cosmos to the microscopic realm of atoms and molecules, from the history of our planet to the depths of the human mind.

"Why do stars shine?" "How did life on Earth begin?" "What is the nature of consciousness?" These are just a few of the questions that have intrigued, puzzled, and inspired generations of thinkers, scientists, and philosophers. In this book, we embark on a quest to seek answers to these questions and many more. We delve into the realms of science, philosophy, history, and culture, weaving together a tapestry of knowledge that paints a clearer picture of how our world truly operates.

The Cosmic Perspective

Our journey begins with the cosmos—the vast and incomprehensible universe that envelops our tiny blue planet. When we gaze upon the night sky, we are beholding the light of distant stars, the remnants of celestial explosions, and the echoes of cosmic events that transpired eons ago. The cosmos is a testament to the grandeur and the sublime beauty of the natural world.

As we delve into the cosmos, we journey through the cosmic timeline. We explore the birth and evolution of stars, the fiery crucibles where elements are forged in the cosmic furnaces. We contemplate the enigmatic nature of galaxies, those vast cosmic cities of stars that dot the cosmic landscape. And we peer into the inky abyss of black holes, where the laws of physics seem to warp and bend.

In our exploration of the cosmos, we are humbled by the vastness of space and the immensity of time. We ponder the mysteries of dark matter and dark energy, the invisible forces that shape the fate of the universe. And we reflect upon the possibility of extraterrestrial life, the tantalizing question of whether we are alone in the cosmos.

The Earth as Our Cradle

As we descend from the cosmic scale to the terrestrial, we arrive at the planet Earth—the crucible of life and the stage upon which the drama of geological forces unfolds. The Earth is not merely a passive backdrop to our lives; it is a dynamic and ever-changing world, shaped by processes that operate on timescales ranging from seconds to millions of years.

To understand the Earth is to understand the forces that have sculpted its surface, to appreciate the ebb and flow of continents, and to witness the majesty of its landscapes. We delve into the science of geology, the discipline that explores the composition, structure, and history of our planet. Geologists are the detectives of the Earth, investigating the rocks, minerals, and fossils that provide clues to its past. They study the forces that shape the landforms we observe today, from the rugged mountains to the meandering rivers.

Central to our understanding of Earth's dynamic nature is the transformative idea of plate tectonics. This theory explains the movement of the Earth's lithosphere, the outermost shell of our planet. It describes the interactions of tectonic plates, massive slabs of rock that float on the semi-fluid asthenosphere beneath them. These plates continually drift, collide, and pull apart, shaping the continents and oceans. Plate tectonics not only explains the formation of mountain ranges and ocean basins but also plays a crucial role in Earth's climate and the distribution of life on our planet.

Life's Story

Our journey continues with life on Earth, from its origins in the primordial soup to the incredible diversity of species that populate our world. The theory of evolution, proposed by Charles Darwin in the 19th century, serves as our guide through the tapestry of life. Evolution is the grand unifying theory that explains the diversity of life on Earth. It posits that species change over time, driven by mechanisms such as natural selection, genetic variation, and reproductive success.

Imagine a population of animals with a range of traits—some tall, some short, some with long beaks, and some with short ones. In an ever-changing environment, some of these traits might provide an advantage for survival. For instance, long-beaked birds may be better equipped to reach food sources in deep flowers. Over generations, individuals with advantageous traits are more likely to survive and reproduce, passing on those traits to their offspring. Gradually, these traits become more common in the population, leading to evolutionary change.

Natural selection, often described as "survival of the fittest," doesn't necessarily mean the strongest or fastest survive. It's about those individuals whose traits best fit their environment. Thus, evolution is a testament to the adaptability and resilience of life on Earth.

As we delve into the story of life, we marvel at the interconnectedness of all living beings. The Tree of Life, a metaphorical representation of the evolutionary history of all species, illustrates the profound truth that all life on Earth shares a common ancestry. The branches of this tree represent the branching and diverging paths that have led to Earth's breathtaking biodiversity. From the tiniest microbe to the largest mammal, every species is connected through this intricate web of life.

The Invisible Forces

Venturing further, we enter the realm of the invisible forces that govern our world at the atomic and subatomic levels. At this scale, the ordinary rules of the physical world seem to break down, and reality takes on a surreal and perplexing character. This is the realm of quantum physics, a field of science that defies common intuition and challenges our fundamental understanding of the universe.

In the quantum world, particles can exist in multiple states simultaneously, seemingly teleport across vast distances, and become entangled in ways that defy classical physics. We explore the mysterious behavior of particles and the fundamental forces that shape the universe. Quantum physics is a realm where the line between reality and the bizarre blurs, where uncertainty reigns, and where the observer plays a fundamental role in shaping the outcome of experiments.

The Human Experience

Finally, we turn our gaze inward to explore the human experience—the realm of consciousness, thought, and emotion. The human mind is a complex tapestry, a labyrinth of neurons and synapses that gives rise to our thoughts, our emotions, and our sense of self. Understanding the mind is a journey of self-discovery, a quest to unravel the intricacies of human cognition and behavior.

The enigma of consciousness, a topic that has perplexed philosophers and scientists for centuries, takes center stage in our exploration. What is consciousness? How does it arise from the activities of the brain? Is it a mere byproduct of physical processes, or does it hold a deeper significance? These are questions that lie at the intersection of science, philosophy, and spirituality.

We delve into the mysteries of the brain, the organ responsible for our thoughts, our memories, and our perceptions of the world. We explore the intricate web of neurons and synapses, the electrochemical dance that underlies our cognitive processes. The human brain is a marvel of evolution, a biological masterpiece that enables us to contemplate the cosmos, create art, and ponder the nature of reality.

The human experience extends beyond the confines of the individual mind to encompass our interactions with society and culture. We reflect on the sweep of human history, from our earliest ancestors to the rise of civilizations and the tapestry of world cultures. We examine the evolution of human societies, the emergence of political systems, and the complex interplay of cultures that have shaped the course of history.

Unraveling Complexity

Throughout our exploration, we aim to unravel the complexity of the world. We strive to bridge the gap between scientific knowledge and everyday understanding, making complex concepts accessible without sacrificing their depth and significance. By the end of this journey, we hope to inspire wonder, ignite curiosity, and provide a richer understanding of "How the World Really Works."

This book is not just a collection of facts and theories; it is an invitation to marvel at the world's wonders, to embrace the beauty of knowledge, and to recognize the profound interconnectedness of all things. It is an acknowledgment that the more we learn, the more there is to discover, and that our quest for understanding is a never-ending voyage.

As we embark on this intellectual odyssey, let us keep our minds open, our questions alive, and our curiosity unquenchable. Together, we will embark on a journey of exploration, enlightenment, and appreciation for the marvelous world that surrounds us. Welcome to "How the World Really Works."

Chapter 1: The Universe Unveiled

1.1 From the Cosmos to Our Cosmic Address

Exploring the Vastness of Space

The universe, with its unfathomable expanse and unending mysteries, has beckoned to humanity since time immemorial. As we embark on this journey through the cosmos, we venture beyond our earthly boundaries to explore the cosmos and its vastness. In this section, "From the Cosmos to Our Cosmic Address," we delve into the cosmic scales, journeying from our humble planet to the outer reaches of the universe, contemplating the intricacies of the cosmos and our place within it.

The Cosmic Canvas

Imagine gazing at the night sky, away from the blinding lights of civilization, and into the vastness of space. The celestial dome above you is a cosmic canvas painted with the shimmering light of distant stars. Each tiny point of light represents a sun, perhaps with its own planets and mysteries. This view, with its captivating beauty and endless potential, has fascinated humanity for millennia.

A Sense of Scale

To truly appreciate the universe, we must first grasp the scale on which it operates. Our cosmic journey begins with an understanding of the astronomical units that allow us to measure these vast distances. The astronomical unit (AU) is the average distance between the Earth and the Sun, approximately 93 million miles (150 million kilometers). It serves as our cosmic yardstick for measuring distances within our solar system.

However, as we venture beyond our solar neighborhood, the distances become mind-boggling.

Consider the speed of light, a fundamental constant of nature traveling at 186,282 miles per second (299,792,458 meters per second). Even at this astonishing speed, light takes about 8 minutes and 20 seconds to travel from the Sun to Earth. This concept alone underscores the vastness of space.

The Solar System: Our Cosmic Neighborhood

Our cosmic address begins with our home: the solar system. This system comprises the Sun, eight planets, their moons, asteroids, comets, and countless other celestial objects. It's a fascinating neighborhood where each planet has its own unique characteristics, from the sweltering surface of Venus to the frigid plains of Pluto.

The Sun, a seething ball of hot, glowing gas, dominates our solar system. It provides the energy that sustains life on Earth and governs the motions of all celestial objects within its gravitational influence. Surrounding the Sun are the planets, each with its own distinct personality.

The Inner Planets

The inner planets, known as the terrestrial planets, are Mercury, Venus, Earth, and Mars. They are characterized by solid, rocky surfaces and relatively small sizes compared to the outer, gas giant planets. Let's take a brief tour of these inner worlds:

Mercury: As the closest planet to the Sun, Mercury experiences extreme temperature variations. Its surface can reach scorching temperatures of up to 800 degrees Fahrenheit (427 degrees Celsius) during the day, while plummeting to -290 degrees Fahrenheit (-180 degrees Celsius) at night.

Venus: Often called Earth's "evil twin," Venus is known for its thick, toxic atmosphere and blistering surface temperatures. The greenhouse effect on this planet traps heat, making it the hottest planet in our solar system.

Earth: Our home planet, Earth, is the only known celestial body to support life. It boasts a diverse biosphere, abundant liquid water, and a range of climates that support a rich variety of ecosystems.

Mars: Often referred to as the "Red Planet" due to its reddish appearance, Mars has long captured the human imagination. It features a thin atmosphere and a landscape marked by vast deserts, towering volcanoes, and a canyon system that dwarfs the Grand Canyon on Earth.

The Outer Planets

Beyond the asteroid belt, we encounter the outer planets, also known as gas giants. These immense worlds—Jupiter, Saturn, Uranus, and Neptune—differ significantly from their terrestrial counterparts:

Jupiter: The largest planet in our solar system, Jupiter is a behemoth of gas and storms. Its massive size and powerful magnetic field make it a cosmic protector, deflecting potential threats like asteroids and comets.

Saturn: Saturn is renowned for its magnificent ring system, which consists of thousands of individual ringlets made of icy particles. These rings create a dazzling spectacle when viewed through a telescope.

Uranus: Unlike most planets, Uranus rotates on its side, giving it a unique appearance and leading to extreme seasonal variations. It's also known for its faint ring system and numerous moons.

Neptune: As the most distant of the gas giants, Neptune is a frigid, windswept world. It's famous for its striking blue color, which is attributed to the presence of methane in its atmosphere.

The Kuiper Belt and Beyond

Beyond Neptune lies the Kuiper Belt, a region populated by icy bodies and dwarf planets, including Pluto. The Kuiper Belt marks the boundary of our solar system, but it is not the final frontier. Far beyond it, the Sun's influence wanes, and we enter the realm of interstellar space.

The Cosmic Address: Earth's Place in the Universe

Our cosmic journey brings us to the concept of our cosmic address—a way to locate our position in the universe. This address is akin to finding our place on a vast cosmic map, helping us understand our relative position among the countless stars and galaxies.

One of the ways to express our cosmic address is through the hierarchical structure of the universe. Let's explore how this structure is organized:

Planet Earth: Our planetary home is the third rock from the Sun, orbiting within the habitable zone where conditions are suitable for life as we know it.

The Solar System: Earth is part of the solar system, a collection of celestial objects gravitationally bound to the Sun. This includes the Sun itself, the eight planets, their moons, asteroids, and comets.

The Milky Way Galaxy: Zooming out further, we discover that our solar system resides within the Milky Way galaxy. The Milky Way is a barred spiral galaxy with an estimated 100 to 400 billion stars, and it is just one of billions of galaxies in the observable universe.

The Local Group: Beyond the Milky Way, we are part of a small galactic neighborhood known as the Local Group. This group comprises more than 54 galaxies, including the Milky Way and the Andromeda Galaxy, our nearest galactic neighbor.

The Observable Universe: Expanding our perspective even

further, we arrive at the observable universe—a mind-boggling expanse filled with galaxies, stars, planets, and cosmic wonders. Our observable universe is estimated to be about 93 billion light-years in diameter.

Cosmic Scales and Distances: To convey the vastness of the universe, astronomers use the unit of the light-year—the distance that light travels in one year at the speed of light. Even at this incredible velocity, cosmic distances are so immense that the light from distant stars takes thousands, millions, or even billions of years to reach us. When we observe the night sky, we are witnessing the universe as it appeared in the distant past.

Cosmic Exploration and Beyond

Our cosmic address is not a fixed point but rather a dynamic one. Humanity's quest for exploration and understanding knows no bounds. From the earliest astronomers who charted the night sky to the modern-day space missions that venture beyond our solar system, the human drive to explore the cosmos continues unabated.

In recent decades, robotic spacecraft have ventured to the

outer planets, sending back invaluable data and breathtaking images of worlds previously beyond our reach. Telescopes both on Earth and in space have peered deeper into the universe, revealing distant galaxies and celestial phenomena that challenge our understanding of the cosmos.

Humanity's desire to explore is not limited to the confines of our solar system. Space agencies around the world have set their sights on missions to Mars, the outer planets, and even the distant Kuiper Belt objects. Beyond that, ambitious plans for human missions to Mars and the search for habitable exoplanets beyond our solar system ignite our imagination and drive our exploration of the cosmos.

The Cosmic Perspective: Awe and Humility

As we conclude our journey from the cosmos to our cosmic address, we are left with a profound sense of awe and humility. The universe is a vast and wondrous place, filled with countless mysteries waiting to be unraveled. Our cosmic address reminds us that we are but tiny specks in the grand tapestry of existence, yet we possess the remarkable ability to contemplate and explore the cosmos.

In the chapters that follow, we will continue to unravel the secrets of the universe, delving deeper into the birth and fate of stars, the intricacies of galaxies, and the enigma of dark matter. Our cosmic journey is far from over, and the wonders of the cosmos await us.

As we venture forth, let us carry with us the knowledge that we are explorers, dreamers, and seekers of truth. The cosmos is our canvas, and the stars are our guideposts. Together, we will navigate the cosmos and uncover the beauty and complexity of "How the World Really Works."

1.2: The Birth and Fate of Stars

In the vast tapestry of the universe, stars occupy a central and mesmerizing role. They are the celestial engines that illuminate the cosmic darkness, shaping the very fabric of the cosmos itself. The story of stars is one of immense gravity—both literal and figurative—where colossal forces and delicate balances converge to give birth to these luminous giants and ultimately dictate their fate.

The Stellar Nursery: Birth of Stars

Our journey into the life of stars begins in the stellar nurseries, vast regions within galaxies where the process of star formation unfolds. At the heart of this process is a cosmic dance between gravity, gas, and dust— a dance that brings stars into existence.

Gravity, the omnipresent force that governs the dynamics of the universe, plays a pivotal role in star formation. Within the cold and dense clouds of gas and dust that populate stellar nurseries, gravity's relentless pull begins to take hold. As regions of higher density within these clouds accumulate, they become gravitational focal points, drawing in more material and steadily increasing in mass.

As the mass of these regions continues to grow, the intense pressure at their cores begins to rise. This escalating pressure is countered by an outward force resulting from the heat generated by gravitational collapse. This delicate balance between gravity's inward pull and thermal pressure's outward push is known as hydrostatic equilibrium. When these forces reach equilibrium, a critical turning point is reached.

Protostars: The Cosmic Embryos

At the heart of this gravitational tug-of-war, a protostar begins to take shape. This protostar is a newborn entity, still in the throes of formation. It is a hot, dense core surrounded by a cocoon of gas and dust that obscures its brilliance from view. This cocoon is often referred to as a protostellar disk, and within its confines, the embryonic star continues to accumulate mass from the surrounding material.

As the protostar gathers more material and the core temperature rises, a pivotal moment approaches When the temperature at the core reaches a critical threshold—around 10 million degrees Celsius (18 million degrees Fahrenheit)—nuclear fusion ignites. Hydrogen atoms at the core fuse together to form helium, releasing an immense amount of energy in the process.

This moment marks the birth of a star—the point at which it begins to shine with its own radiant light. The newborn star is now known as a main-sequence star, and it embarks on a journey of stellar adulthood, converting hydrogen into helium through the process of nuclear fusion. This fusion process not only powers the star but also generates the energy that radiates outward, illuminating the cosmos.

The Life of a Star: A Delicate Balance

Once a star is born, its fate is inexorably linked to the delicate balance between gravity's inexorable pull and the energy generated by nuclear fusion in its core. This balance determines the star's size, temperature, luminosity, and ultimately, its destiny.

Main-sequence stars, like our Sun, spend the majority of their lives stably fusing hydrogen into helium in their cores. This phase of a star's life is marked by equilibrium, where the inward gravitational force is counteracted by the outward pressure generated by the fusion reactions in the core. For stars like the Sun, this stable phase can last billions of years.

The specific path a star takes through its life cycle depends on its mass. More massive stars burn through their hydrogen fuel more quickly and consequently have shorter lifespans. On the other hand, less massive stars, like red dwarfs, can shine steadily for trillions of years.

As a star exhausts its hydrogen fuel, the balance of forces within its core begins to shift. The core contracts, causing the outer layers to expand and the star to swell in size. This marks the transition to the next phase in a star's life cycle.

The Stellar Swan Song: Supernovae and Beyond

In the later stages of a star's life, when it has exhausted its nuclear fuel, the core undergoes dramatic changes. For stars with mass greater than about 1.4 times that of the Sun, the core's contraction leads to an explosive event of cataclysmic proportions—a supernova.

A supernova is one of the most energetic and luminous events in the universe. In a matter of seconds, a dying star can outshine an entire galaxy. The explosion disperses heavy elements forged in the star's core throughout space, seeding the cosmos with the raw materials necessary for the formation of planets, and, ultimately, life.

The remnants of a supernova can take different forms. In some cases, they collapse into incredibly dense objects known as neutron stars, where matter is packed so tightly that a teaspoonful would weigh as much as a mountain. In other cases, the core collapse continues, forming a black hole—an enigmatic cosmic entity with gravity so intense that not even light can escape its grasp.

The End of the Stellar Odyssey

As stars spend their nuclear fuel, they go through various stages, from main-sequence stars to red giants, and for the most massive, end their lives in explosive supernovae. The remnants they leave behind can become neutron stars or black holes, contributing to the ongoing drama of the cosmos.

For stars like our Sun, the final chapter is a more serene one. As it runs out of hydrogen to burn in its core, it will expand into a red giant, shedding its outer layers into space, creating beautiful cosmic nebulae. Eventually, the core will cool and contract, becoming a white dwarf—a dense, Earth-sized remnant that will slowly fade over billions of years.

The birth and fate of stars are not just cosmic events but integral chapters in the grand narrative of the universe. Stars, in their various forms and life stages, shape the galaxies they inhabit, influence the formation of planetary systems, and provide the elements necessary for life as we know it.

As we contemplate the birth and demise of stars, we gain a deeper understanding of the cosmos and our place within it. The universe, with its celestial wonders and intricate processes, continues to inspire awe and curiosity, beckoning us to explore the mysteries that lie beyond the boundaries of our planet.

Stellar Afterlife: White Dwarfs and Beyond

For stars like our Sun, the final act in their cosmic journey is marked by the transformation into a white dwarf. A white dwarf is an extraordinary celestial object—small in size, yet incredibly dense. The core of the once-glowing star contracts and cools, leaving behind a remnant that is about the size of Earth but with a mass comparable to the Sun.

White dwarfs are not active nuclear reactors like their parent stars; instead, they radiate heat and gradually fade away over billions of years. This slow cooling process leads to a remarkable evolution. Initially, a white dwarf shines brightly, but as it loses energy, it transitions from a white-hot glow to a cooler, reddish hue. Eventually, it becomes a cold and dark "black dwarf."

In the vast expanse of the cosmos, the journey of white dwarfs does not end with their transformation into black dwarfs. The universe itself is dynamic and subject to change, and over timescales that dwarf human comprehension, even black dwarfs may not remain static.

They could eventually be influenced by external forces, such as nearby supernovae or encounters with other celestial objects, causing them to undergo further transformation.

Stellar Recycling: Supernova Remnants and Nebulae

The aftermath of a supernova explosion is a spectacle of cosmic proportions. The immense energy released in a supernova blast scatters elements forged within the star's core into the surrounding space. This process plays a crucial role in the enrichment of the cosmos, as it distributes heavy elements like carbon, oxygen, and iron, which are essential for the formation of planets, life, and even ourselves.

Supernova remnants are the glowing, expanding shells of gas and dust left behind by the explosion. These remnants continue to interact with their surroundings, influencing the dynamics of interstellar gas clouds and fostering the birth of new stars. In a way, the death of a star becomes the catalyst for the birth of others, perpetuating the cycle of stellar life.

Furthermore, the dispersal of material by supernovae contributes to the creation of stunning cosmic nebulae. These vast, glowing clouds of gas and dust are often illuminated by nearby stars or the remnants of the supernova itself. Nebulae are celestial artistry, capturing the imagination with their intricate shapes and vivid colors. Some of the most famous examples include the Eagle Nebula and the Orion Nebula, both showcasing the ongoing interplay between birth and death in the cosmos.

The Ultimate Fate: Black Holes and the Fabric of Space-time

For the most massive stars, those with a mass several times that of our Sun, the final act is not a graceful transition into a white dwarf but a dramatic collapse into a black hole. Black holes are perhaps the most enigmatic and perplexing objects in the universe, where the very fabric of space and time is warped beyond recognition.

When a massive star's core can no longer withstand gravitational collapse, it implodes, forming a singularity—an infinitesimally dense point where gravity becomes infinitely strong. Surrounding this singularity is the event horizon, a boundary beyond which nothing, not even light, can escape. This is the point of no return, where the laws of physics as we know them break down.

The formation of a black hole is a cosmic catastrophe, and it marks the end of the star's journey. However, black holes are not merely cosmic vacuum cleaners that swallow everything in their path. They are also powerful engines, capable of generating intense gravitational forces and, in some cases, emitting powerful jets of energy and radiation.

Black holes come in various sizes, from stellar-mass black holes formed by the collapse of massive stars to supermassive black holes residing at the centers of galaxies, including our own Milky Way. These supermassive black holes can have masses equivalent to millions or even billions of times that of the Sun, and their influence extends far beyond their event horizons, shaping the dynamics of entire galaxies.

The Cosmic Continuum: From Birth to Rebirth

The birth and fate of stars are integral components of the cosmic continuum—a never-ending cycle of creation and destruction that defines the universe's tapestry. Stars, in their myriad forms, are not isolated entities but interconnected participants in the cosmic drama.

As we delve into the mysteries of star birth and death, we gain insights into the profound interconnectedness of the cosmos. The elements forged within stars become the building blocks of planets, and in the fertile grounds of planetary systems, life itself takes root. Our existence, as well as our ability to ponder the universe, owes much to the celestial processes that have unfolded over billions of years.

The study of stars also extends our understanding of fundamental physics and the nature of space and time. Black holes, with their gravitational extremes, challenge our comprehension of the universe's fabric. They remind us that the cosmos is a realm of both wonder and enigma, where even the most exotic phenomena are governed by the laws of physics.

In contemplating the birth and fate of stars, we embark on a journey through time and space, from the incandescent birth of a protostar to the mysterious depths of a black hole's singularity. The universe's story is written in the life cycles of stars, and as we peer into the night sky, we are witness to the ongoing cosmic narrative—a story that continues to unfold, beckoning us to explore its secrets and embrace its profound beauty.

Stellar Nurseries: Cosmic Cradles of Creation

To truly appreciate the birth of stars, we must delve deeper into the wondrous realms of stellar nurseries. These colossal clouds of gas and dust are the cosmic cradles of star formation, where the intricate dance of matter and energy unfolds.

Within these stellar nurseries, temperatures plummet to frigid extremes, and dense pockets of gas and dust begin to coalesce under the inexorable pull of gravity. As these regions accumulate more material, their gravitational attraction intensifies, drawing in even more matter from their surroundings.

Imagine standing amidst a swirling cosmic dust storm, where countless tiny particles are drawn together, bonding through gravitational attraction. Over millions of years, these particles merge, forming clumps that continue to grow as they sweep up more material. The gravitational pressure at their cores mounts, setting the stage for the birth of a star.

Protostars: The Cosmic Embryos

As the gravitational pressure at the core of these gas and dust clumps escalates, temperatures rise, and a protostar emerges from the cosmic haze. This protostar is an infant in the celestial sense, a radiant heart cocooned within a shroud of gas and dust, still in the throes of formation.

The protostar continues to feed on the surrounding material, growing in size and mass. It resembles a cosmic incubator, nurturing the burgeoning star-to-be. This phase can be likened to a celestial adolescence, as the protostar matures and its core temperature steadily climbs.

The key milestone marking the transition from protostar to bona fide star is the ignition of nuclear fusion at the core. When the core temperature reaches around 10 million degrees Celsius

(18 million degrees Fahrenheit), the hydrogen atoms in the core fuse together to form helium, releasing a torrent of energy in the process. This radiant energy marks the star's birth, and it begins to shine with its own radiant light.

Main-Sequence Stars: The Stellar Adulthood

Once a star has ignited its nuclear fusion engine, it enters the main-sequence phase—a period of relative stellar stability where the balance between gravity's inward pull and the outward pressure generated by nuclear fusion maintains equilibrium.

Main-sequence stars come in various sizes, from the petite red dwarfs to the sun-like yellow dwarfs, and even massive blue giants. The specific path a star takes within this phase depends on its mass, with more massive stars burning through their hydrogen fuel more quickly than their smaller counterparts.

Our own Sun, a yellow dwarf, falls into this category. It has been shining steadily for billions of years and will continue to do so for several billion more. As a main-sequence star, the Sun steadily converts hydrogen into helium at its core, releasing a steady stream of energy that illuminates our solar system and sustains life on Earth.

Stellar Evolution: The Bigger They Are...

For more massive stars, the main-sequence phase is a relatively brief chapter in their stellar evolution. The intense gravitational pressure at their cores, combined with their greater mass, causes them to burn through their hydrogen fuel at an accelerated pace. This rapid consumption of nuclear fuel ultimately leads to more dramatic and fiery destinies.

Massive stars can evolve into red super giants, where their outer layers expand dramatically.

During this phase, they can swell to sizes that dwarf the entire orbit of Earth around the Sun. These colossal stars exhibit a fiery beauty, their radiance visible across vast cosmic distances.

However, this spectacular phase is fleeting. The end of the line for massive stars is marked by a cataclysmic explosion—the supernova. This stupendous event rivals the radiance of an entire galaxy for a brief moment, scattering heavy elements into space and heralding the star's transformation into something altogether different.

Supernovae: Cosmic Fireworks

The death knell of a massive star reverberates through the cosmos with the fury of a supernova. This explosive event is among the most energetic phenomena in the universe, outshining entire galaxies for a short time.

The supernova explosion is the result of the star's core collapsing under the relentless force of gravity, followed by a dramatic rebound caused by the intense heat generated during core collapse. This rebound triggers a shockwave that races outward, tearing through the star's outer layers and catapulting them into space at velocities approaching a fraction of the speed of light.

The material ejected by a supernova is rich in heavy elements, including those necessary for the formation of planets, life, and the very essence of our existence. As the remnants of a supernova disperse into space, they leave behind a stunning cosmic tapestry, a testament to the universe's propensity for both creation and destruction.

Remnants and Remembrances: Neutron Stars and Black Holes

The aftermath of a supernova is a stark reminder of the cosmic cycle of birth and death. For some massive stars, the core collapse leads to the formation of a neutron star, an incredibly dense and rapidly rotating object. Neutron stars are cosmic marvels, where matter is compressed to densities beyond comprehension. A teaspoonful of neutron star material would weigh as much as a mountain on Earth.

In other cases, the core collapse continues unabated, giving rise to a black hole—an entity so enigmatic and mysterious that it defies our conventional understanding of space, time, and gravity. Black holes are regions where the gravitational pull is so intense that nothing, not even light itself, can escape their grasp.

These exotic objects are both captivating and mystifying, challenging our understanding of the universe's fundamental principles. Black holes come in different sizes, from stellar-mass black holes to supermassive ones that reside at the centers of galaxies, including our Milky Way.

Continuity and Change: The Cosmic Dance of Stars

The birth and fate of stars are integral components of the cosmic narrative—a never-ending dance of matter and energy, creation and destruction, equilibrium and cataclysm. Stars, whether they are the fiery giants or the enigmatic remnants, are threads in the intricate tapestry of the universe.

The elements forged within stars—carbon, oxygen, iron, and others—are the building blocks of planets, life, and the cosmos itself. We, as inhabitants of a small blue planet orbiting an unassuming yellow dwarf star, owe our existence to the cosmic processes that have unfolded over billions of years.

The study of stars not only deepens our understanding of the universe but also enriches our comprehension of fundamental physics, the nature of space and time, and the intricate connections that bind the cosmos together.

As we gaze upon the night sky, we are witnesses to the ongoing drama of creation and destruction, a story that continues to unfold in the boundless expanse of the cosmos. The birth and fate of stars are not isolated events but chapters in the grand narrative of the universe, inviting us to explore its mysteries and embrace its profound beauty.

1.3: Galaxies, Black Holes, and Dark Matter

In our quest to unravel the mysteries of the universe, we've journeyed from the birth and fate of individual stars to the grand stage where countless stars congregate—galaxies. In this chapter, we embark on a cosmic odyssey to explore the intricate tapestry of galaxies, the enigmatic realms of black holes, and the elusive nature of dark matter.

Galaxies: Cosmic Island Cities

In the vast and enigmatic expanse of the cosmos, galaxies stand as some of the most captivating and majestic entities. They are the cosmic island cities, each a unique assemblage of stars, gas, dust, and dark matter, separated by the immense voids of intergalactic space.

In this exploration, we delve deep into the mesmerizing world of galaxies, uncovering their diversity, significance, and their pivotal role in shaping our understanding of the universe.

Astronomical Islands: The Cosmic Abodes of Stars

Imagine for a moment that you are standing on a serene beach, gazing out at a vast ocean. On the horizon, a cluster of islands emerges, each distinct in its shape, size, and character. These islands, surrounded by the boundless sea, are reminiscent of galaxies in the cosmic landscape.

Galaxies are celestial islands adrift in the cosmic ocean. They are immense conglomerations of stars, gas, dust, and dark matter bound together by gravity. Each galaxy is a realm of its own, housing anywhere from millions to hundreds of billions of stars. These stars, like the inhabitants of an island city, go about their lives in the vast expanse of their galactic home.

Our own Milky Way galaxy is one such island city in the cosmos. It spans approximately 100,000 light-years in diameter and contains billions of stars, including our Sun. Viewed from afar, the Milky Way presents a graceful spiral structure with spiral arms winding outward from a central bar. It is a place of wonder, where starbirth and stardust coexist in cosmic harmony.

Diversity of Galaxies: The Many Faces of Cosmic Cities

Just as Earth boasts a diversity of landscapes, from towering mountains to sprawling plains, the universe presents a myriad of galaxies, each with its own character and allure. Galaxies come in various shapes and sizes, forming a rich tapestry of cosmic diversity.

Spiral Galaxies: Spirals are perhaps the most iconic of all galaxies. They are characterized by their elegant, swirling arms that wrap around a bright central nucleus. The Milky Way is a spiral galaxy, and so is our neighboring Andromeda Galaxy. These galaxies exude grace and symmetry, and their spiral arms are often hotbeds of star formation.

Elliptical Galaxies: In stark contrast to spirals, elliptical galaxies are more like cosmic orbs. They lack the spiral structure and appear as smooth, elongated ellipses. Elliptical galaxies come in various sizes, from giants with millions of stars to dwarfs with only a fraction of that number. They are often composed of older stars and contain less interstellar gas.

Irregular Galaxies: As the name suggests, irregular galaxies defy convention. They lack the symmetry and structure of spirals and ellipticals, often appearing as chaotic and asymmetrical collections of stars and gas. Irregular galaxies are a testament to the unpredictable nature of the cosmos.

Dwarf Galaxies: Dwarf galaxies are the smallest cosmic islands. They can be either spiral, elliptical, or irregular in shape. Dwarf galaxies are numerous, and they often accompany larger galaxies like the Milky Way as satellites. Studying dwarf galaxies is essential for understanding the dynamics of galactic neighborhoods.

Lenticular Galaxies: Lenticular galaxies combine features of both spirals and ellipticals. They have a central bulge like ellipticals but also possess a flattened disk, similar to spirals. These galaxies are often in a transitional phase, perhaps evolving from one type into another.

Interacting and Merging Galaxies: Sometimes, galaxies engage in celestial dance. Their gravitational interactions can result in breathtaking displays of cosmic fireworks as they collide and merge. Such interactions can trigger intense bursts of star formation and lead to the formation of new galaxies.

Active Galaxies: Some galaxies exhibit heightened levels of activity, often emanating from their central regions. These active galactic nuclei (AGN) can include powerful phenomena like quasars, which are among the most luminous objects in the universe. AGN are fueled by supermassive black holes at their cores.

Galactic Histories: Chronicles of Starbirth and Stardust

Galaxies are not static entities but dynamic realms where the cosmic cycle of starbirth and stardust continues unabated. They are cosmic chronicles, preserving the history of their stars and the processes that have shaped their existence.

Star Formation: Within galaxies, vast clouds of gas and dust serve as the stellar nurseries where new stars are born. The gravitational collapse of these clouds initiates the formation of protostars, which eventually ignite nuclear fusion at their cores, marking their entry into the main sequence phase. The energy radiated by these stars illuminates the galaxy's landscape.

Stellar Evolution: Galaxies are filled with stars of various ages and life stages. Just like our Sun, stars within galaxies undergo a life cycle that includes stages like red giants, white dwarfs, and supernovae. Supernovae, in particular, play a critical role in enriching galaxies with heavy elements like carbon, oxygen, and iron, which are essential for the formation of planets and life.

Galactic Dynamics: The interactions between stars, as well as the gravitational influence of dark matter, shape the overall dynamics of a galaxy. Spiral galaxies, for example, maintain their graceful arms through the gravitational pull of surrounding stars. In elliptical galaxies, the motions of stars are more random, reflecting their older stellar populations.

Galactic Neighborhoods: Galaxies do not exist in isolation; they are part of larger cosmic neighborhoods. Clusters of galaxies, called galaxy clusters, form vast cosmic cities where galaxies are drawn together by gravity. These clusters are surrounded by cosmic voids, immense expanses of empty space where galaxies are scarce.

The Milky Way: Our Galactic Home

For us, Earth-bound inhabitants of the cosmos, the Milky Way holds a special significance. It is our home, our cosmic island city in the immense sea of the universe. The Milky Way's central bar and spiral arms have guided human navigation for millennia, as they stretch across the night sky, inviting our contemplation.

As we peer into the night sky, we are gazing into the heart of our galaxy, and the stars we see are just a fraction of the billions that call the Milky Way home. Our Sun is but one of these stars, and it is orbited by a retinue of planets, including Earth, the cradle of humanity.

The Milky Way's history is interwoven with our own, and its evolution has played a pivotal role in shaping the conditions for life on Earth. The elements that form our bodies and the planets around us were forged within the cores of stars in the Milky Way. We are, quite literally, made of stardust.

The Cosmic Quest: Understanding the Universe through Galaxies

Studying galaxies is not merely an exercise in celestial tourism; it is a fundamental pursuit of cosmic understanding. Galaxies offer us windows into the universe's past, present, and future. They reveal the processes of star formation, the dynamics of cosmic neighborhoods, and the influence of dark matter.

Galaxies are cosmic laboratories, allowing astronomers to test theories of stellar evolution, galaxy formation, and the nature of dark matter. They also serve as cosmic yardsticks, enabling us to measure distances in the universe and probe the expansion of the cosmos itself.

In the quest to understand the universe, galaxies are like pages in a grand cosmic storybook. Each one tells a tale of stellar birth and death, of interactions with neighboring galaxies, and of the relentless march of cosmic time. As we turn these pages, we uncover the secrets of the universe's evolution and its enduring mysteries.

The Future of Galactic Exploration

The study of galaxies continues to be a dynamic field of research, driven by advances in technology and observational capabilities. Cutting-edge telescopes, both on Earth and in space, provide unprecedented views of distant galaxies and their inner workings. Observatories like the Hubble Space Telescope have revolutionized our understanding of the cosmos.

Future missions and telescopes, such as the James Webb Space Telescope, promise to expand our knowledge even further, peering deeper into the universe's history and shedding light on the nature of the earliest galaxies that formed in the cosmos.

Moreover, the study of galaxies extends beyond traditional astronomy. It intersects with other fields like astrophysics, cosmology, and particle physics, offering interdisciplinary opportunities for exploration and discovery.

In Conclusion: Cosmic Tapestry of Galaxies

Galaxies are not merely objects of astronomical study; they are the cosmic tapestry woven from the threads of stars, gas, dust, and dark matter. Each galaxy tells a story of cosmic evolution and holds the secrets of the universe's past and future.

As we gaze upon the night sky and contemplate the beauty of galaxies, we are reminded of our place in the grand cosmic narrative. We are inhabitants of a small blue planet orbiting an unassuming star in one of billions of galaxies. Yet, our curiosity and quest for understanding have allowed us to peer into the depths of the universe, unlocking its wonders and complexities.

In the cosmic island cities of galaxies, we find both familiarity and mystery, unity and diversity. They are the cosmic beacons that guide our exploration of the universe, inviting us to unravel their secrets and embrace the profound interconnectedness of all things.

As we continue our journey through the cosmos, galaxies remain as beacons of inspiration, reminding us of the boundless wonders that await our discovery in the eternal sea of the universe.

Supermassive Black Holes: Cosmic Monsters

In the profound and mysterious theater of the universe, supermassive black holes stand as some of the most enigmatic and awe-inspiring entities. These colossal cosmic monsters, lurking at the heart of many galaxies, defy our conventional understanding of reality. In this exploration, we plunge into the shadowy realms of supermassive black holes, unraveling their secrets, significance, and the profound influence they exert on the cosmic stage.

The Cosmic Abyss: Introduction to Supermassive Black Holes

Imagine, if you will, a region of space where gravity is so extreme that nothing, not even light, can escape its clutches. This is the eerie domain of black holes, regions where the fabric of spacetime is so warped that they become cosmic abysses. Among black holes, a distinct class of giants reigns supreme—the supermassive black holes.

Supermassive black holes are the behemoths of the black hole family, with masses that can range from millions to billions of times that of our Sun. These cosmic monsters reside at the centers of many galaxies, including our own Milky Way. Their sheer size and insatiable appetite for matter make them objects of fascination and terror in the cosmic landscape.

The Birth of Supermassive Black Holes: Cosmic Seeds

The origin of supermassive black holes remains a topic of active research and debate among astronomers and astrophysicists. The prevailing theory suggests that these colossal entities may have formed through a combination of processes, including:

Accretion: Supermassive black holes likely began as smaller black holes, formed through the gravitational collapse of massive stars. Over time, these black holes accreted surrounding matter, growing in size and mass. The relentless pull of gravity drew in stars, gas, dust, and other cosmic debris, feeding the black hole's insatiable hunger.

Galactic Mergers: Galaxy collisions and mergers play a pivotal role in the formation and growth of supermassive black holes. When two galaxies collide, their central black holes can eventually merge into a larger supermassive black hole. The resulting black hole can have a mass greater than the sum of its predecessors.

Quasar Activity: Some of the most luminous and energetic

objects in the universe are quasars, which are powered by supermassive black holes at their centers. These quasars emit intense radiation and are thought to represent a phase in the evolution of supermassive black holes when they are actively accreting matter.

The Cosmic Monarchs: Central to Galactic Evolution

Supermassive black holes hold a pivotal role in shaping the evolution and destiny of galaxies. They are not passive cosmic entities but active participants in the galactic drama. Here are some ways in which they exert their influence:

Galactic Structure: Supermassive black holes help define the structure of their host galaxies. They are often found at the center of galaxies, where they play a role in regulating the distribution of stars and gas. In spiral galaxies like the Milky Way, the central black hole can stabilize the rotation curve and prevent stars from being flung into intergalactic space.

Star Formation Regulation: The intense radiation and energy emitted by supermassive black holes can influence star formation within their host galaxies. They can heat and ionize surrounding gas, potentially quenching star formation in certain regions. Conversely, they can trigger starbursts in other regions through their gravitational interactions.

Galactic Dynamics: The presence of a supermassive black hole can lead to intricate dynamics within a galaxy. Stars and other objects orbit the central black hole, creating a complex dance of celestial bodies. This dynamic interaction can be observed through the study of stellar orbits and the measurement of the black hole's gravitational influence.

Active Galactic Nuclei (AGN): Supermassive black holes at the centers of galaxies can exhibit active behavior, generating powerful phenomena known as active galactic nuclei (AGN). These phenomena can include quasars, blazars, and Seyfert galaxies, which are characterized by intense emissions of energy and radiation. AGN can profoundly affect the galactic environment and the behavior of matter within their vicinity.

The Cosmic Engine: Supermassive Black Holes and Quasars

One of the most extraordinary manifestations of supermassive black holes' activity is the phenomenon known as quasars. Quasars are among the brightest and most energetic objects in the universe, outshining entire galaxies. They are powered by the gravitational pull of supermassive black holes at their cores.

The central regions of quasars contain a swirling accretion disk of hot gas and matter spiraling into the supermassive black hole. As this matter is drawn into the black hole's gravitational grasp, it releases a tremendous amount of energy in the form of radiation. The intense brightness of quasars is a result of this energy release.

Quasars serve as cosmic beacons, visible across vast

cosmic distances. They provide astronomers with a unique opportunity to study the universe's early epochs, as the light from distant quasars has traveled billions of years to reach us. This allows scientists to probe the conditions of the universe in its infancy.

Galactic Cannibalism: The Role of Supermassive Black Holes in Mergers

Galaxy collisions and mergers are common occurrences in the universe's history. When two galaxies converge, the gravitational interactions between them can lead to the formation of a single, larger galaxy. In this process, the central supermassive black holes of each galaxy are drawn together, eventually merging into an even more massive black hole.

This cosmic cannibalism has significant implications for the evolution of galaxies and their central black holes:

Black Hole Growth: The merger of two galaxies brings their central supermassive black holes into close proximity. As they spiral toward each other under the influence of gravity, they release gravitational waves—ripples in spacetime predicted by Einstein's theory of general relativity. The release of these waves leads to a coalescence of the black holes, resulting in a single, more massive black hole.

Energy Release: The merger process is not quiet; it is accompanied by the release of intense energy and radiation. This energy can have far-reaching effects on the surrounding galactic environment. It can heat and expel gas from the central regions of the newly formed galaxy, potentially quenching star formation.

Feedback Mechanism: The energy released during black hole mergers can serve as a feedback mechanism, influencing the future growth of galaxies and their central black holes. It can regulate the rate of star formation and the buildup of matter around the black hole.

The Dark Heart of the Milky Way: Sagittarius A and Beyond*

At the heart of our Milky Way galaxy lies a cosmic enigma known as Sagittarius A*, or Sgr A*. It is the supermassive black hole that presides over our galactic neighborhood, located approximately 26,000 light-years from Earth. While Sgr A* is comparatively modest in size compared to some other supermassive black holes, it plays a crucial role in the dynamics of the Milky Way.

Despite its significance, Sgr A* is notoriously difficult to observe directly. It resides in the densest region of the galaxy, surrounded by a veil of gas and dust that obscures its visibility in visible and optical wavelengths of light. However, astronomers have employed innovative techniques and instruments, such as radio and infrared observations, to peer into the heart of our galaxy.

Recent discoveries have revealed the dynamic nature of Sgr A*. It experiences occasional flares of energy and matter as it devours nearby gas and dust. The study of these flares provides valuable insights into the behavior of supermassive black holes and their interactions with their surroundings.

Cosmic Mysteries and Future Exploration

While we have made significant strides in understanding supermassive black holes, they remain shrouded in mystery. Some of the key questions and areas of ongoing research include:

The Nature of Black Holes: The true nature of black holes, including the structure of their event horizons and the behavior of matter within them, continues to be a subject of intense investigation. The study of black hole mergers and gravitational waves offers a glimpse into these enigmatic objects.

Black Hole Growth: Understanding the mechanisms that fuel the growth of supermassive black holes, including the accretion of matter and the role of mergers, is a key area of research. Studying the early universe and the formation of the first black holes provides insights into their origins.

Feedback Effects: Exploring the feedback effects of supermassive black holes on their host galaxies is essential for understanding galactic evolution. How do black holes regulate star formation and influence the galactic environment?

Cosmic Evolution: Supermassive black holes are intimately tied to the evolution of galaxies and the large-scale structure of the universe. Studying their distribution across cosmic time provides a window into the universe's history.

Gravitational Waves: The detection of gravitational waves from black hole mergers has opened a new era of astrophysical exploration. Ongoing efforts to detect and study these waves offer a unique opportunity to probe the fundamental nature of black holes.

The Role of Dark Matter: The connection between supermassive black holes and dark matter—the elusive substance that makes up a significant portion of the universe's mass—remains a subject of investigation. How does dark matter influence the formation and dynamics of galaxies and their central black holes?

The Cosmic Monarchs and Human Curiosity

Supermassive black holes, these cosmic monsters, stand as

cosmic monarchs, reigning over their galactic domains. They challenge our understanding of the universe and beckon us to explore the furthest reaches of the cosmos.

As we delve into the mysteries of supermassive black holes, we are reminded of the profound interconnectedness of all things in the universe. From the smallest particles to the largest cosmic giants, the cosmos is a tapestry of complexity and wonder.

Human curiosity knows no bounds, and our relentless pursuit of knowledge has led us to the brink of understanding these cosmic enigmas. With each discovery, we peel back another layer of the universe's secrets, revealing the grandeur and intricacy of the cosmic stage.

In the cosmic realm of supermassive black holes, we encounter the extremes of gravity, energy, and spacetime curvature. These cosmic monsters challenge our perceptions and ignite our imagination. They are cosmic beacons, guiding us to explore the very limits of our understanding.

Dark Matter: The Hidden Maestro

In the grand symphony of the cosmos, dark matter plays the role of an elusive but essential conductor. This hidden maestro, imperceptible to our senses, orchestrates the movements of galaxies, shapes the large-scale structure of the universe, and leaves its indelible mark on the cosmic stage. In this exploration, we embark on a journey to unveil the mysteries of dark matter, delving into its nature, influence, and profound significance in our quest to understand the universe.

The Cosmic Puzzle: The Need for Dark Matter

Our understanding of the universe is built upon the foundations of gravity and motion, as famously described by Sir Isaac Newton's laws of motion and universal gravitation. According to these principles, the gravitational attraction between massive objects governs the motions of celestial bodies, including planets orbiting stars and stars within galaxies.

However, when astronomers turned their telescopes toward galaxies in the early 20th century, they made a perplexing discovery. The observed motions of stars within galaxies did not conform to the predictions based on the visible matter alone. In fact, stars at the outskirts of galaxies were moving at nearly the same velocities as those near the galactic center, contrary to what classical physics would suggest.

This discrepancy presented a cosmic puzzle: where was the missing mass that exerted the gravitational pull necessary to explain these observed motions? The visible matter, such as stars, gas, and dust, could not account for the observed gravitational forces. It became clear that something invisible and elusive was at play—an enigmatic substance we now refer to as dark matter.

The Nature of Dark Matter: An Enigma Unveiled

Dark matter, as its name implies, does not emit, absorb, or interact with electromagnetic radiation like light or other forms of radiation. It is, in essence, invisible to our traditional methods of detection. This elusiveness has made the identification of dark matter a formidable challenge for physicists and astrophysicists.

To unravel the nature of dark matter, scientists have employed a variety of approaches, each offering a piece of the puzzle:

Gravitational Effects: Dark matter's presence is inferred

through its gravitational effects on visible matter. The gravitational pull of dark matter helps galaxies maintain their structural integrity, prevents them from flying apart as they rotate, and shapes their large-scale distribution in the universe.

Cosmic Microwave Background: The cosmic microwave background radiation, a remnant of the early universe, provides clues about the composition of the cosmos. Measurements of the cosmic microwave background suggest that dark matter makes up a significant portion of the universe's total mass-energy content.

Gravitational Lensing: The phenomenon of gravitational lensing occurs when the gravitational field of dark matter bends and distorts the paths of light from distant objects. Observations of gravitational lensing provide indirect evidence for the presence of dark matter, as they reveal the mass distribution in galaxy clusters and other cosmic structures.

Particle Physics: The search for dark matter particles represents a frontier of particle physics. Theoretical models propose a variety of candidate particles, such as weakly interacting massive particles (WIMPs) and axions, which could make up dark matter. Experiments are underway to detect these particles directly or indirectly.

Simulations: Numerical simulations of the universe's evolution, incorporating dark matter's gravitational effects, have been instrumental in our understanding of cosmic structure formation. These simulations enable scientists to compare theoretical predictions with observed data, refining our knowledge of dark matter's distribution.

The Cosmic Gravemaker: Dark Matter's Gravitational Influence

While dark matter may remain hidden from our senses, its

gravitational influence is unmistakable and profound. Here are some of the key ways in which dark matter shapes the cosmos:

Galactic Stability: Dark matter acts as a cosmic glue, binding galaxies together. It provides the additional gravitational mass needed to counterbalance the gravitational pull of visible matter, such as stars and gas. Without dark matter, galaxies would be significantly less stable and might disintegrate.

Galactic Rotation Curves: Dark matter plays a crucial role in explaining the flat rotation curves observed in galaxies.

According to classical physics, the orbital velocity of stars should decrease with increasing distance from the galactic center. However, observations reveal that stars in the outer regions of galaxies move at nearly constant velocities, indicating the presence of unseen dark matter.

Large-Scale Structure: On the grandest scales, dark matter acts as a cosmic scaffold, shaping the large-scale structure of the universe. It forms the invisible framework upon which galaxies and galaxy clusters are woven, creating a vast cosmic web of filaments and voids.

Cosmic Microwave Background: The cosmic microwave background radiation, a snapshot of the universe's early state, provides evidence for dark matter's gravitational effects. Variations in the temperature of this radiation across the sky are attributed to the density fluctuations in the early universe, which were influenced by the presence of dark matter.

Galaxy Cluster Dynamics: Dark matter dominates the mass of galaxy clusters, which are the largest gravitationally bound structures in the universe. Its gravitational pull on galaxies within clusters causes them to move at high velocities and contributes to the overall mass distribution of the cluster.

The Dark Matter Enigma: Candidate Particles and Detection Efforts

The nature of dark matter remains one of the most profound mysteries in modern physics. Scientists have proposed a multitude of candidate particles that could make up dark matter, each with its own unique properties and behaviors. Some of the leading candidates include:

Weakly Interacting Massive Particles (WIMPs): WIMPs are hypothetical particles that interact through weak nuclear force and gravity. They are considered one of the most compelling dark matter candidates due to their potential for relatively high mass and weak interactions with ordinary matter. Numerous experiments, such as the Large Underground Xenon (LUX) and the Cryogenic Dark Matter Search (CDMS), have been designed to detect WIMPs directly.

Axions: Axions are light, neutral particles that were originally proposed to solve a problem in particle physics related to the strong nuclear force. They are another candidate for dark matter, although they are exceptionally difficult to detect due to their low mass and feeble interactions.

Sterile Neutrinos: Neutrinos are elusive subatomic particles that interact very weakly with matter. Sterile neutrinos are hypothetical heavier cousins of standard neutrinos and are considered as potential dark matter candidates. Experiments like the IceCube Neutrino Observatory aim to detect sterile neutrinos indirectly.

MACHOs and WIMPs: In addition to particle candidates, dark matter may also consist of massive compact halo objects (MACHOs), such as black holes or dense stellar remnants, and primordial black holes. Although these objects are less favored as dark matter candidates, they remain part of the ongoing investigation.

Axion-Like Particles (ALPs): ALPs are hypothetical particles related to axions that could potentially make up dark matter. Experimental searches for ALPs involve observing how they interact with magnetic fields.

Efforts to detect and identify dark matter particles are ongoing, involving a combination of laboratory experiments, astronomical observations, and theoretical modeling. The discovery of dark matter particles would represent a profound breakthrough in our understanding of the universe's composition and structure.

The Cosmic Enigma and Cosmic Conundrums

Dark matter remains a cosmic enigma, challenging our understanding of the universe and the fundamental forces that govern it. As we continue our exploration of this hidden maestro, we encounter a series of cosmic conundrums and profound questions:

Dark Matter's True Nature: What is the true nature of dark matter? Is it composed of one or more particles, and what are their properties? Answering these questions is central to unraveling the dark matter mystery.

Dark Matter and Particle Physics: How does dark matter interact with ordinary matter, if at all? Could dark matter particles reveal themselves through their interactions with known particles in the laboratory?

Dark Matter and Cosmic Structure: How did dark matter influence the formation of galaxies, galaxy clusters, and the cosmic web? How does it relate to the large-scale structure of the universe?

Cosmic Web and Filaments: How do dark matter's gravitational effects shape the intricate web of cosmic filaments and voids that make up the large-scale structure of the universe?

Dark Matter and Galaxy Evolution: How has dark matter influenced the evolution of galaxies over cosmic time? How does it impact star formation and the dynamics of galactic systems?

Cosmic Connections: What is the connection between dark matter and other cosmic mysteries, such as the nature of dark energy and the ultimate fate of the universe?

The Cosmic Maestro's Final Bow: The Quest Continues

In the cosmic symphony, dark matter remains a hidden maestro, conducting the cosmic movements from the shadows. As we strive to unveil its nature and significance, we are reminded of the boundless wonders and mysteries of the universe.

The quest to understand dark matter is a journey of both scientific exploration and philosophical contemplation. It challenges us to redefine our understanding of the cosmos, confront the limitations of our knowledge, and push the boundaries of human curiosity and ingenuity.

As we continue our cosmic odyssey, we are drawn deeper into the enigmatic realms of dark matter, inspired by the belief that the universe's most profound secrets are within our reach. With each discovery and each unanswered question, we are beckoned to follow the invisible conductor's baton and to explore the cosmic symphony in its entirety.

The final bow of the cosmic maestro has yet to be taken. The quest to unveil the hidden mysteries of dark matter continues, resonating with the enduring human spirit of curiosity and discovery. In the end, it is the journey itself that enriches our understanding of the universe and leaves us with a profound appreciation for the cosmic enigmas that surround us.

Chapter 2: The Solar System and Planetary Mysteries

2.1 A Journey through Our Solar Neighborhood:

In the vast expanse of the cosmos, our solar system stands as a familiar and intriguing corner of the universe. Comprising the Sun, eight major planets, numerous moons, asteroids, comets, and other celestial objects, our solar system offers a captivating journey through space and time. In this comprehensive exploration, we embark on a voyage through our solar neighborhood, immersing ourselves in the celestial bodies that make up this cosmic community and gaining a profound appreciation for the beauty and complexity of our planetary system.

The Sun: Our Radiant Star

Our journey commences with the radiant and awe-inspiring Sun, the gravitational anchor around which our entire solar system revolves. The Sun is not just a distant celestial body; it is the very source of life on Earth. Its significance in our lives extends far beyond its role as a mere astronomical object. The Sun is the ultimate cosmic engine that sustains life as we know it.

The Anatomy of the Sun

The Sun is a colossal ball of superheated gas, primarily composed of hydrogen (about 74%) and helium (about 24%). Its structure is a fascinating interplay of extreme conditions:

Core: Deep within the Sun's core, temperatures soar to about 15 million degrees Celsius (27 million degrees Fahrenheit), and immense pressures facilitate nuclear fusion. Here, hydrogen atoms collide and fuse into helium, releasing an astonishing amount of energy in the process. This energy production—approximately 4.2 million tons of matter converted into energy every second—fuels the Sun's radiant brilliance.

Radiative Zone: Surrounding the core is the radiative zone, where energy generated in the core gradually radiates outward through the form of photons. The photons, moving in a random walk pattern, can take thousands to millions of years to traverse this region before reaching the next layer.

Convective Zone: Beyond the radiative zone lies the convective zone, where heat is transported by the churning motion of material. Hot plasma rises toward the surface, releasing energy in the form of visible light and heat. Cooler material sinks back into the depths to complete the cycle.

Photosphere: The visible surface of the Sun, known as the photosphere, is the layer from which sunlight emanates. It is a seething cauldron of plasma, and its temperature hovers around 5,500 degrees Celsius (9,932 degrees Fahrenheit). The photosphere is the layer where we observe features like sunspots, which are temporary dark spots caused by intense magnetic activity.

Atmosphere: Above the photosphere, the Sun's atmosphere consists of several distinct layers, including the chromosphere and the corona. These regions exhibit fascinating phenomena, such as solar flares and coronal mass ejections, which have a significant impact on space weather and our technological infrastructure.

The Sun's Energy: A Lifeline for Earth

The Sun's continuous nuclear fusion processes result in the emission of an astonishing amount of energy. This energy, radiated in various forms of electromagnetic radiation, including visible light, ultraviolet light, and X-rays, floods our solar system, bathing the planets and other celestial objects in its life-giving glow.

Earth, positioned at an optimal distance from the Sun, receives just the right amount of solar energy to maintain a habitable climate. The Sun's warmth sustains liquid water on our planet's surface, a fundamental ingredient for life as we know it. Furthermore, photosynthesis in plants and algae harnesses solar energy to convert carbon dioxide and water into oxygen and sugars, laying the foundation of Earth's diverse ecosystems.

Solar radiation also plays a critical role in Earth's climate system. The absorption and reflection of sunlight by our planet's surface and atmosphere drive weather patterns, ocean currents, and the distribution of heat around the globe.

The Solar Wind and the Heliosphere

The Sun's influence extends far beyond its visible surface. The solar corona, the outermost layer of the Sun's atmosphere, gives rise to the solar wind—an outward flow of charged particles, primarily electrons and protons, that permeates our solar system. This solar wind carries with it the Sun's magnetic field, forming what is known as the interplanetary magnetic field (IMF).

As the solar wind interacts with the magnetospheres of planets, it creates dynamic and complex interactions. Earth's magnetosphere, for example, protects our planet from the full brunt of the solar wind, deflecting and channeling charged particles toward the poles, where they produce stunning auroras.

The region beyond the influence of the solar wind is known as the heliosphere—a vast, bubble-like region in space created by the solar wind's outward flow. The Voyager 1 and 2 spacecraft have ventured into this uncharted territory, providing invaluable data about the heliosphere's boundaries and the nature of the interstellar medium beyond.

The Planets: Our Cosmic Companions

As we journey outward from the Sun, we encounter the eight

major planets that orbit it. These planets, diverse in size, composition, and characteristics, offer a kaleidoscope of celestial wonders:

Inner Planets (Terrestrial Planets):

Mercury: Our journey through the solar system begins with the innermost planet, Mercury. Named after the fleet-footed messenger of Roman mythology, Mercury is a rocky world characterized by extreme temperature variations. Its surface, covered in craters and marked by rugged terrain, provides a glimpse into the harsh realities of space.

Venus: Often referred to as Earth's "sister planet" due to its similar size and composition, Venus hides a tumultuous and inhospitable environment beneath its thick cloud cover. A runaway greenhouse effect has rendered its surface incredibly hot, with temperatures soaring to around 467 degrees Celsius (872 degrees Fahrenheit).

Earth: The third planet from the Sun, Earth, stands as a testament to the intricate balance of conditions necessary for life to flourish. Its dynamic geology, teeming oceans, and life-sustaining atmosphere create a nurturing environment that supports a rich tapestry of ecosystems.

Mars: Known as the "Red Planet" due to its rusty appearance, Mars has been a focus of exploration and fascination for centuries. The possibility of liquid water in its past and the presence of subsurface ice have fueled speculation about its potential for harboring life.

These terrestrial planets are characterized by their solid surfaces, relatively small sizes, and proximity to the Sun. Earth, in particular, is a world of immense complexity, where the interplay of geology, climate, and biology has given rise to a stunning diversity of life forms.

Outer Planets (Gas Giants):

Jupiter: The largest of all the planets in our solar system, Jupiter is a colossus with a magnetic personality. Its colossal size and powerful magnetosphere make it a dominant presence. Jupiter's mesmerizing cloud bands, the Great Red Spot, and an entourage of moons and rings have fascinated astronomers for generations.

Saturn: Saturn, renowned for its magnificent ring system, is a gas giant that exemplifies celestial splendor. Its iconic rings, composed of countless particles of ice and rock, present an otherworldly spectacle. Saturn's dynamic atmosphere and diverse moon system have piqued human curiosity for centuries.

Uranus: Uranus, an ice giant, stands out for its peculiar sideways rotation. Its axis of rotation is tilted at a nearly 90-degree angle, causing it to roll through space like a cosmic ball. Its pale blue color and faint ring system add to its enigmatic charm.

Neptune: As the farthest major planet from the Sun, Neptune is a distant and mysterious ice giant. Its deep blue hue, fast atmospheric winds, and the ever-elusive Great Dark Spot contribute to its captivating features.

The outer planets, or gas giants, are fundamentally different from their terrestrial counterparts. They are primarily composed of hydrogen and helium and lack solid surfaces. Instead, their atmospheres transition smoothly into their interiors, which become denser and hotter with depth.

Mysteries of the Solar System: Unanswered Questions

While our understanding of the Sun and the planets in our solar system has grown significantly, the cosmic neighborhood remains a realm filled with mysteries and unanswered questions. These enigmas continue to intrigue scientists and space enthusiasts, inspiring further exploration and discovery:

The Formation of the Solar System: The question of how the solar system formed from a vast cloud of gas and dust, and what processes led to the formation of the Sun and planets, remains one of the fundamental mysteries of astrophysics.

The Search for Life: Mars, with its tantalizing hints of liquid water in the past, continues to be a focal point in the search for extraterrestrial life. Recent missions, such as the Mars rovers and the detection of organic molecules, have fueled optimism about the potential for microbial life or signs of past life.

Planetary Dynamics: The gas giants, particularly Jupiter and Saturn, are dynamic worlds with extreme weather patterns and unique features. Understanding the mechanics of their atmospheres and the origins of their iconic storms, like the Great Red Spot, is an ongoing challenge.

Exploration Beyond: Humanity's aspirations to explore and

eventually colonize other celestial bodies within our solar system, such as the Moon and Mars, raise questions about the technological and logistical challenges that must be overcome to make these ambitions a reality.

The Kuiper Belt and Oort Cloud: Beyond the orbit of Neptune lies the Kuiper Belt, a region inhabited by a diverse array of icy objects, including Pluto. Additionally, the hypothetical Oort Cloud is thought to be a vast, distant reservoir of comets. These regions hold valuable clues about the early solar system and the dynamics of the outer solar system.

As we journey through our solar neighborhood, these mysteries and more beckon us to explore, discover, and better understand the cosmic wonders that surround us. Our solar system is a testament to the beauty and complexity of the universe, and each celestial body within it holds a unique story waiting to be told.

2.2 The Wonders of Our Planetary Neighbors:

As we continue our cosmic journey through our solar system, we shift our focus to the planets themselves. In Section 2.2: "The Wonders of Our Planetary Neighbors," we encounter a diverse cast of celestial bodies, each with its own unique features, landscapes, and mysteries. From the fiery and tempestuous surface of Venus to the frigid and enigmatic world of Pluto, our planetary neighbors offer a captivating tapestry of cosmic wonders.

Venus: Earth's Fiery Twin

Our exploration of the planets takes us first to Venus, often referred to as Earth's "sister planet." While Venus shares several similarities with our home planet, including its size and rocky composition, it also possesses some of the most extreme conditions in our solar system.

Venus's Harsh Environment

Extreme Heat: Venus's thick and suffocating atmosphere traps heat through a runaway greenhouse effect. Surface temperatures soar to an astonishing 467 degrees Celsius (872 degrees Fahrenheit), making it the hottest planet in our solar system—even hotter than Mercury, which is closer to the Sun.

Atmospheric Pressure: The atmospheric pressure on Venus is a crushing 92 times that of Earth's, equivalent to the pressure found at a depth of nearly 3,000 feet (900 meters) in Earth's oceans. This pressure would instantly crush most human-made spacecraft.

Thick Cloud Cover: Venus is shrouded in a thick layer of clouds composed mainly of sulfuric acid. These clouds reflect sunlight, causing the planet to appear brilliantly bright in the sky.

The Venusian Landscape

Despite its hellish conditions, Venus has a strikingly beautiful and otherworldly landscape:

Volcanoes: Venus is home to a multitude of volcanoes, including some of the largest in the solar system. These volcanoes, although typically not actively erupting, bear witness to the planet's tumultuous geological history.

Highland Plateaus: Vast highland plateaus known as "tesserae" stretch across Venus's surface. These regions exhibit complex and fractured terrain, hinting at geological activity and possible tectonic forces.

Venusian Plains: The planet also features extensive plains, some of which are relatively young and smooth, suggesting recent volcanic activity or resurfacing.

Mysteries of Venus

While Venus has been the subject of numerous spacecraft missions, such as NASA's Magellan and the Soviet Union's Venera program, many questions about this enigmatic world remain unanswered:

Runaway Greenhouse Effect: What exactly triggered Venus's extreme greenhouse effect, causing the planet to become a searing inferno? Understanding this phenomenon has implications for our study of climate change on Earth.

Volcanic Activity: Are there currently active volcanoes on Venus, and what drives the planet's extensive volcanic features? Studying Venus's geology can provide insights into the inner workings of terrestrial planets.

Past Oceans: Some scientists speculate that Venus may have had oceans in its distant past. What happened to these potential bodies of water, and could they have once supported life?

Mars: The Red Planet Beckons

Our next planetary neighbor, Mars, has captivated human imagination for centuries. Often referred to as the "Red Planet" due to its distinctive rusty appearance, Mars has been a focus of exploration and scientific inquiry.

Mars's Varied Terrain

Olympus Mons: Mars boasts Olympus Mons, the largest volcano in the solar system, towering at over 69,000 feet (21.9 kilometers). Its enormous size and relatively young surface suggest that Mars may still be volcanically active.

Valles Marineris: A vast canyon system known as Valles Marineris stretches across the Martian landscape, dwarfing the Grand Canyon on Earth. This colossal rift extends over 2,500 miles (4,000 kilometers) in length.

Polar Ice Caps: Mars's polar ice caps are composed of water and carbon dioxide ice. They undergo seasonal changes, expanding and contracting with the changing Martian seasons.

The Search for Martian Life

One of the most compelling aspects of Mars is the possibility of past or present life. Several missions have sought to uncover clues about the planet's potential habitability:

Water on Mars: Evidence of liquid water in Mars's distant past, including river valleys and lake beds, suggests that conditions may have been suitable for life.

Mars Rovers: NASA's rovers, such as Curiosity and Perseverance, are equipped with scientific instruments to study the planet's geology and search for signs of ancient microbial life.

Underground Reservoirs: Recent discoveries of subsurface lakes and reservoirs raise intriguing questions about the potential habitability of Mars beneath its barren surface.

Mysteries of Mars

As our exploration of Mars continues, several mysteries and unanswered questions persist:

Fate of Ancient Oceans: What happened to Mars's ancient oceans, and could they have harbored life in the planet's distant past?

Atmospheric Loss: How did Mars lose its once-thicker atmosphere, and what implications does this have for the planet's climate and potential habitability?

Potential for Life: Is life, either past or present, lurking beneath the Martian surface or in subsurface aquifers? Discovering microbial life on Mars would be a profound scientific breakthrough.

The Outer Giants: Jupiter and Saturn

As we journey further into the outer reaches of our solar system, we encounter the majestic gas giants—Jupiter and Saturn. These colossal planets, with their mesmerizing atmospheres and captivating moons, continue to astound scientists and space enthusiasts alike.

Jupiter: The King of Planets

The Great Red Spot: Jupiter's most iconic feature, the Great Red Spot, has long been a subject of fascination. This colossal storm, which has been observed for centuries, is a swirling vortex of gas with a striking reddish hue. Its exact origin and the reasons behind its enduring nature remain intriguing puzzles. Recent observations suggest that the Great Red Spot may be shrinking, sparking debates about its future.

Atmospheric Layers: Jupiter's atmosphere is a complex and dynamic system composed primarily of hydrogen and helium, along with traces of other gases and compounds. The planet's atmosphere is divided into bands and zones characterized by distinct cloud patterns and colors. The alternating bands of gas move in opposite directions, creating a striking visual display.

Magnetosphere and Auroras: Jupiter boasts one of the most powerful magnetic fields in the solar system. Its magnetosphere extends far beyond the planet, creating a radiation environment that is hazardous to spacecraft. This magnetic field interacts with charged particles from the solar wind, giving rise to spectacular auroras near Jupiter's poles.

Galilean Moons: Jupiter's moon system is a treasure trove of celestial wonders. The four largest moons—Io, Europa, Ganymede, and Callisto—known as the Galilean moons, were discovered by Galileo Galilei in 1610. Each of these moons is a world in its own right, with unique characteristics.

Io: Io is the most volcanically active body in the solar system, with plumes of sulfur dioxide and sulfur erupting from its surface. The intense tidal forces from Jupiter and its sister moons contribute to Io's geological activity.

Europa: Europa is believed to have a subsurface ocean beneath its icy crust, making it a prime target in the search for extraterrestrial life. The presence of water and potential geothermal activity under the ice has sparked interest in exploring Europa further.

Ganymede: Ganymede is the largest moon in the solar system and possesses its own magnetic field. It has a diverse landscape, with both older, heavily cratered regions and younger, smoother terrains indicating past geological activity.

Callisto: Callisto is a heavily cratered moon with a relatively untouched surface. Its surface features provide insights into the history of impacts in the outer solar system.

NASA's Juno Mission: The Juno spacecraft, which arrived at Jupiter in 2016, is on a mission to study the planet's interior, gravity field, magnetic field, and polar magnetosphere. Juno's findings are shedding light on Jupiter's structure and the processes driving its magnetic field.

Saturn: The Ringed Wonder

Stunning Ring System: Saturn's most renowned feature is its magnificent ring system, consisting of countless particles of ice and rock. These rings extend hundreds of thousands of kilometers from the planet and are divided into several distinct rings, including the prominent A,

B, and C rings. The origin of Saturn's rings remains a subject of study, with hypotheses ranging from the capture of a moon to the breakup of a larger moon.

Shepherd Moons: Saturn's ring dynamics are influenced by "shepherd moons" like Pan and Daphnis. These small moons gravitationally interact with the ring particles, sculpting and maintaining the ring structures. The interactions between these moons and the rings create intricate patterns and divisions.

Titan: A Moon with an Atmosphere: Saturn's largest moon, Titan, is a world of particular interest. It is the only moon in the solar system with a substantial atmosphere, primarily composed of nitrogen. Titan's thick, hazy atmosphere conceals its surface from visible light, but radar imaging by the Cassini spacecraft revealed vast lakes and rivers of liquid methane and ethane.

Enceladus: Geysers of Life? Enceladus, another of Saturn's moons, is known for its geysers of water ice erupting from its south pole. These geysers contain organic molecules, suggesting the presence of subsurface oceans and raising the possibility of habitable environments.

Cassini-Huygens Mission: The Cassini spacecraft, which orbited Saturn from 2004 to 2017, provided an unprecedented view of Saturn, its rings, and its moons. The Huygens probe even descended to the surface of Titan, becoming the first human-made object to land on a moon in the outer solar system.

Uranus and Neptune: The Ice Giants

As we venture further into the outer realms of our solar system, we encounter Uranus and Neptune, the enigmatic ice giants. These distant planets present a stark contrast to their gas giant cousins, Jupiter and Saturn, with their unique characteristics and mysteries that continue to pique scientific curiosity.

Uranus: The Tilted Giant

Unusual Axial Tilt: One of the most distinctive features of Uranus is its extreme axial tilt. Unlike most planets in our solar system, which have relatively small axial tilts, Uranus appears to roll on its side, with an axial tilt of nearly 90 degrees. This peculiar orientation remains one of the great mysteries of our cosmic neighborhood.

Moons and Rings: Uranus has a collection of moons, the largest of which are Titania and Oberon. Additionally, it possesses a faint and complex ring system. These moons and rings provide valuable insights into the planet's history and dynamics. The Voyager 2 spacecraft conducted a flyby of Uranus in 1986, capturing detailed images and data about the planet and its surroundings.

Unique Magnetosphere: Uranus's magnetic field is peculiar in that it is tilted relative to its rotation axis. This misalignment between the magnetic field and the planet's axis of rotation is thought to be related to Uranus's unusual tilt. The dynamics of Uranus's magnetosphere are a subject of ongoing study.

Neptune: The Distant Blue Giant

Deep Blue Hue: Neptune, known as the "Distant Blue Giant," derives its striking color from the presence of methane in its atmosphere, which absorbs red light. Its atmosphere is marked by fast-moving storms and features, including the ever-elusive Great Dark Spot, reminiscent of Jupiter's Great Red Spot.

Triton: A Captive Moon: Neptune's moon Triton is of particular interest due to its retrograde orbit, which suggests that it may have been captured by Neptune's gravity. Triton's surface features geysers that spew nitrogen gas into space, creating an atmosphere and raising questions about the moon's geological activity.

Exploring the Ice Giants: While the Voyager 2 spacecraft provided valuable data about Uranus and Neptune during its flybys, these ice giants remain relatively unexplored compared to their gas giant counterparts. Future missions to these distant worlds hold the potential to uncover more about their atmospheres, interiors, and moons.

Mysteries of the Ice Giants

Both Uranus and Neptune pose intriguing questions and mysteries for scientists to unravel:

Uranus's Axial Tilt: What caused Uranus's extreme axial tilt, and how does it impact the planet's climate and magnetic field? Understanding this unique feature could provide insights into the planet's history.

Neptune's Great Dark Spot: The dynamics of Neptune's Great Dark Spot and its similarity to Jupiter's Great Red Spot continue to puzzle scientists. Studying these storm systems can shed light on the behavior of gas giants' atmospheres.

Composition and Interior: Further exploration of Uranus and Neptune's interiors could reveal the composition of these ice giants, including the presence of water, ammonia, and methane. Understanding their internal structures is key to unraveling their history and evolution.

Moons and Rings: The moons and rings of Uranus and Neptune offer windows into the planet's past and gravitational interactions. Studying their origins and characteristics can provide valuable insights into the dynamics of the outer solar system.

Pluto: The Dwarf Planet Beyond

Beyond the orbit of Neptune lies the distant and enigmatic Pluto, a celestial object that has captivated the imagination of astronomers and space enthusiasts for decades. Although it was once considered the ninth planet in our solar system, Pluto has since been reclassified as a dwarf planet, a change that sparked scientific discussions and led to a more precise understanding of our cosmic neighborhood.

New Horizons Mission: Pluto's recent fame was reignited by NASA's New Horizons spacecraft, which conducted a historic flyby of Pluto in 2015. This flyby provided humanity with its first close-up views of this distant world, unveiling a wealth of geological diversity and intriguing features.

Surface Features: New Horizons revealed a landscape on Pluto that is far more complex than expected. It unveiled towering ice mountains that soar as high as 11,000 feet (3,500 meters) and vast plains of frozen nitrogen. The heart-shaped feature, informally known as "Tombaugh Regio" after Pluto's discoverer, Clyde Tombaugh, became an iconic image.

Charon and Pluto's Moon System: Pluto is accompanied by a system of five known moons, with Charon being the largest. Charon is nearly half the size of Pluto itself, making their relationship unique in the solar system. The study of Pluto's moon system provides insights into the formation and history of this distant world.

Kuiper Belt and Beyond: Pluto's location in the Kuiper Belt, a region beyond the orbit of Neptune, is part of a vast population of small icy bodies. The Kuiper Belt is a treasure trove of objects that preserve the early history of our solar system, and it continues to be a target for exploration and discovery.

Beyond Pluto: As New Horizons ventured deeper into the Kuiper Belt, it conducted a flyby of another Kuiper Belt object named Arrokoth in 2019. This object's unique shape and pristine condition provided valuable insights into the early solar system's formation.

Dwarf Planet Status: The reclassification of Pluto as a dwarf planet by the International Astronomical Union (IAU) in 2006 sparked debates about the criteria for planetary status. While Pluto may no longer be considered a full-fledged planet, it remains an object of scientific interest and public fascination.

Future Exploration: The exploration of Pluto and the Kuiper Belt continues to be a focus of scientific research and future missions. Proposed missions aim to study more Kuiper Belt objects and unveil the mysteries of this distant region of our solar system.

2.3 Mysteries of the Solar System: Unveiling the Cosmic Enigmas

Our solar system, a remarkable cosmic arena composed of planets, moons, asteroids, and comets, has captivated humanity's imagination for centuries. As we gaze at the night sky and observe the celestial bodies that adorn it, we are drawn into a world of wonder and intrigue. The solar system is a testament to the awe-inspiring forces of nature and the mysteries that persist, challenging our understanding of the cosmos. In this section, we embark on a journey to explore the enduring mysteries that beckon scientists, astronomers, and space enthusiasts to delve deeper into the unknown.

Origins of the Solar System: The Birth of Cosmic Complexity

At the heart of our solar system's mysteries lies its very origin. How did this complex system of planets, moons, and other celestial objects come into existence from a primordial cloud of gas and dust? The story of the solar system's formation is a tale of cosmic transformation that has fascinated scientists and astronomers for centuries.

The leading theory of solar system formation is the solar nebula hypothesis. According to this model, our solar system began as a massive cloud of gas and dust known as a molecular cloud. Under the influence of gravity, this cloud collapsed, forming a spinning disk. In the center of this disk, the Sun ignited, becoming the radiant heart of our solar system.

Meanwhile, within the spinning disk, dust particles began to stick together, forming small clumps called planetesimals. These planetesimals continued to collide and merge, eventually giving rise to the planets we know today. However, this process raises numerous questions and mysteries:

Short Timescales: One mystery is related to the timescales involved. The formation of planets from dust particles within the disk appears to be a relatively rapid process in astronomical terms, happening within millions of years. Understanding how these processes could occur so quickly challenges our current models.

Composition Variations: The diversity of planets in our solar system, from the rocky worlds like Earth to the gas giants like Jupiter, raises questions about how the composition of the solar nebula varied across the disk. Why did some regions of the disk favor the formation of gas giants while others gave rise to terrestrial planets?

Solar System Anomalies: The peculiarities in the solar system, such as the alignment of the planets' orbits, the presence of dwarf planets, and the abundance of asteroids and comets, continue to intrigue scientists. These features hint at a complex and dynamic formation process.

Planet Nine: The Elusive Wanderer

In the outer realms of the solar system, a captivating mystery awaits discovery. Astronomers have long suspected the presence of an undiscovered ninth planet, often referred to as "Planet Nine" or "Planet X." This hypothetical planet is believed to reside in the distant reaches of the Kuiper Belt, a region beyond the orbit of Neptune, and its existence is inferred from gravitational anomalies observed in the movements of distant celestial objects.

The key mystery surrounding Planet Nine is its exact nature and characteristics. While its gravitational influence on Kuiper Belt objects suggests a substantial mass, its size, composition, and other attributes remain unknown. Several hypotheses have been proposed, including the possibility that it could be a gas giant or a massive, icy world. However, the hunt for Planet Nine continues, with astronomers scouring the night sky in search of this enigmatic wanderer.

The Oceans of Europa and Enceladus: Hidden Habitats for Life?

The search for extraterrestrial life within our solar system has been a driving force behind space exploration. Two of the most promising locations for potential life are the moons Europa, which orbits Jupiter, and Enceladus, which orbits Saturn. These moons are believed to harbor subsurface oceans beneath their icy exteriors, raising tantalizing questions about the potential for life beyond Earth.

Europa's Hidden Ocean: Europa, one of Jupiter's Galilean moons, is of particular interest due to its subsurface ocean. This vast, liquid water ocean lies beneath a relatively thin layer of ice. Heat generated by tidal interactions with Jupiter's gravity is thought to keep the ocean from freezing solid. The presence of liquid water and potential geothermal activity beneath the ice makes Europa a prime candidate in the search for extraterrestrial life.

Enceladus: Geysers of Organic Molecules: Saturn's moon Enceladus has gained attention for its geysers of water ice and organic molecules erupting from its south pole. The presence of complex organic compounds in these plumes suggests the possibility of habitable environments below the moon's icy shell.

Both Europa and Enceladus pose profound mysteries:

Habitability: Are these subsurface oceans habitable? Could they harbor life, even in the absence of sunlight? Understanding the conditions within these hidden oceans and the potential for life is a central focus of future exploration.

Origin of Oceans: How did these moons acquire their subsurface oceans, and how have these oceans evolved over time? Investigating the origin and history of these hidden waters is a critical scientific endeavor.

Mars: The Red Planet's Ancient Secrets

Mars, often referred to as the "Red Planet," has long held a place of fascination in the realm of planetary exploration. It bears a striking resemblance to Earth in many ways and has fueled the imagination with the possibility of past or present life. The mysteries of Mars extend to its ancient history and the potential clues it may hold regarding the conditions for life beyond our planet.

Evidence of Liquid Water: Mars exhibits numerous features that suggest the presence of liquid water in its distant past. River valleys, ancient lake beds, and mineral deposits all provide evidence of a once-watery world. The mystery lies in understanding the duration and stability of these ancient water bodies and their potential implications for the existence of life.

Methane on Mars: The detection of methane in Mars's atmosphere has raised intriguing questions. On Earth, methane is often produced by living organisms. The presence of methane on Mars could point to either geological processes or the possibility of microbial life beneath the surface.

Exploration and Habitability: Mars exploration missions, including rovers like Curiosity and Perseverance, seek to uncover more about the planet's past habitability and the potential for life. The quest to find evidence of past microbial life on Mars continues to be a central scientific objective.

Venus: From Hellish Inferno to Potential Habitability

Venus, often referred to as Earth's "sister planet," presents a stark contrast between its potential habitability and its current hostile environment. While Venus's surface conditions are akin to a hellish inferno, its history hints at the possibility of a more Earth-like past. The mysteries of Venus lie in understanding its climatic evolution and whether it could have once been a more hospitable world.

Runaway Greenhouse Effect: Venus's extreme surface temperatures are the result of a runaway greenhouse effect. Its thick atmosphere, primarily composed of carbon dioxide, traps heat, leading to scorching temperatures. Understanding the factors that led to this runaway greenhouse effect and whether it could have been averted in the past is a central question.

Past Liquid Water: Evidence from radar mapping missions suggests that Venus may have had liquid water on its surface in the distant past. The presence of valleys, channels, and other geologic features indicates the possibility of a more temperate climate in the past. Investigating this history is a scientific challenge.

Habitability: A Venus Connection?: Could Venus, in its early history, have been habitable? Understanding the processes that transformed it into the inhospitable world we see today sheds light on the broader question of habitability in the cosmos.

The Rings of Saturn: A Magnificent Enigma

Saturn's stunning ring system is one of the most iconic features in our solar system. Composed of countless particles of ice and rock, these intricate rings have fascinated astronomers and space enthusiasts for centuries. While we have gained considerable knowledge about Saturn's rings, they continue to present mysteries.

Ring Formation: The exact formation mechanisms of Saturn's rings are not fully understood. Various hypotheses suggest they could be remnants of a shattered moon, captured material, or the result of a long and complex evolutionary process.

Ring Age and Stability: Determining the age and stability of Saturn's rings is a challenge. Are they relatively young and dynamic, or have they persisted for billions of years? Understanding the longevity of these exquisite rings is a topic of ongoing research.

Shepherd Moons: Saturn's ring dynamics are influenced by "shepherd moons" like Pan and Daphnis. These small moons gravitationally interact with the ring particles, sculpting and maintaining the ring structures. The intricate dance between these moons and the rings is a complex puzzle.

Mercury's Exosphere: A Thin Veil of Mystery

Mercury, the closest planet to the Sun, is a world of extremes.

Its surface experiences dramatic temperature variations, from scorching heat to frigid cold. Yet, one of the mysteries of Mercury lies in its tenuous exosphere, a wispy and tenuous atmosphere-like layer that shrouds the planet.

Exosphere Composition: Mercury's exosphere is primarily composed of elements such as hydrogen, helium, and oxygen. Understanding the sources of these exospheric constituents and how they behave in the planet's harsh environment is a topic of ongoing research.

Solar Wind Interaction: Mercury's proximity to the Sun exposes it to intense solar radiation and the solar wind. The interplay between the solar wind and Mercury's exosphere raises questions about how this interaction affects the planet's surface and exospheric dynamics.

Volatiles and Surface Interaction: The exosphere contains traces of volatile elements like sodium and potassium. Investigating how these volatile elements interact with Mercury's surface and escape into space is essential for unraveling the planet's history.

The Kuiper Belt and Beyond: Beyond Pluto's Realm

Beyond the orbit of Pluto lies a vast and largely unexplored region known as the Kuiper Belt.

This region is populated with a diverse array of icy bodies, including dwarf planets, asteroids, and comets. The Kuiper Belt offers a treasure trove of clues about the early solar system's formation and evolution.

Kuiper Belt Object Diversity: The variety of objects within the Kuiper Belt is striking, ranging from Pluto, the largest known Kuiper Belt object, to smaller icy bodies like Haumea and Makemake. Understanding the origins and dynamics of this population is a scientific endeavor.

Undiscovered Dwarf Planets: While Pluto is the most famous of the Kuiper Belt objects, many more dwarf planets likely await discovery. Characterizing their properties, orbits, and compositions is an ongoing challenge.

Exploration and Sample Return: Future missions to Kuiper Belt objects hold the potential to bring back samples from these distant worlds. Analyzing these samples can provide critical insights into the early solar system's composition and history.

The Role of Cosmic Collisions: Shapers of Planetary Destinies

Throughout the solar system's history, cosmic collisions have played a central role in shaping its destiny. Impact events, involving asteroids, comets, and other celestial bodies, have left their mark on planetary surfaces and have had far-reaching consequences.

Frequency of Impacts: Determining the frequency of impact events, both past and future, is a critical scientific endeavor. Understanding the risk of potentially hazardous impacts on Earth and other celestial bodies is vital for planetary defense.

Impact Consequences: The consequences of impact events, from the formation of craters to mass extinctions, have shaped the course of planetary evolution. Investigating the geological and biological effects of impacts is a multidisciplinary challenge.

Planetary Evolution: Cosmic collisions have influenced the geological evolution of planets and moons. Studying the scars left by impacts provides valuable insights into the history of celestial bodies.

The Fate of the Sun: A Celestial Timepiece

The Sun, our radiant star, is central to the existence of life on Earth. However, like all stars, it has a finite lifespan, and its eventual fate is a cosmic mystery. The Sun's future evolution holds profound implications for the solar system and its inhabitants.

Stellar Evolution: The Sun is currently in the main sequence phase of its life, where it fuses hydrogen into helium. However, in the distant future, it will exhaust its hydrogen fuel and undergo changes, including expansion into a red giant.

Solar System Consequences: The Sun's transformation into a red giant will have profound consequences for the solar system. It will likely expand, engulfing the inner planets, including Earth. Understanding these future changes is part of the study of stellar evolution.

End of the Sun's Life: The Sun's ultimate fate involves its transition from a red giant to a white dwarf—a dense, Earth-sized remnant. The processes leading to this endpoint and the potential influence on the solar system's remnants are subjects of scientific investigation.

Conclusion: Embracing the Cosmic Enigma

The mysteries of our solar system, from its enigmatic origins to the fate of the Sun, form a captivating narrative that transcends time and space. Each unanswered question invites us to embark on a journey of discovery, to peer into the cosmic unknown, and to unveil the secrets that shape our celestial neighborhood.

As we explore the enduring mysteries of the solar system, we embrace the spirit of inquiry and the pursuit of knowledge that drive scientific exploration. The cosmos, with its boundless wonders and enigmatic complexities, remains an ever-unfolding story—one that continues to inspire awe, curiosity, and a deep sense of wonder.

In the chapters to come, we will venture even further into the mysteries of the universe, delving into the quantum realm, exploring the intricacies of the human mind, and embracing the wonders of our cosmic neighborhood and beyond. The cosmic enigma beckons us to journey onward, to seek understanding, and to revel in the profound mysteries that define our existence in the cosmos.

Chapter 3: The Earth and Its Systems

3.1: Geology and Plate Tectonics

Our world, the Earth, is a complex and ever-evolving planet. The science of geology, which studies the Earth's structure, composition, and the processes that shape it, allows us to unravel the mysteries of our planet's history and its ongoing transformation. In this chapter, we embark on a journey deep into the heart of the Earth's geology and delve into the groundbreaking theory of plate tectonics.

The Earth's Layers: A Multilayered Puzzle

To understand geology and plate tectonics, we must first grasp the intricate structure of our planet. The Earth can be divided into several distinct layers, each with its unique characteristics and properties.

The Crust: We start at the outermost layer, the Earth's crust, the solid ground we walk on. The crust is relatively thin compared to the other layers, and it is where geological processes like erosion, volcanism, and mountain-building primarily occur. Despite its thinness, the crust is extraordinarily diverse, with various rock types and landforms.

The Mantle: Beneath the crust lies the Earth's mantle, an extensive layer composed of semi-solid rock. The mantle extends hundreds of kilometers beneath the Earth's surface and is where we find the source of heat that drives many geological processes. The movement of material in the mantle plays a critical role in plate tectonics.

The Outer Core: Continuing our descent, we encounter the Earth's outer core, a layer composed of liquid iron and nickel. The outer core's movement generates the Earth's magnetic field through the geodynamo process, which is crucial for our planet's protection from solar radiation.

The Inner Core: At the very heart of the Earth lies the inner core, a solid sphere primarily composed of iron and nickel. Despite the immense heat, pressure, and density, the inner core remains solid due to the tremendous pressure at this depth. Temperatures here can reach several thousand degrees Celsius.

Understanding the Earth's layered structure is essential to comprehend how geological processes operate and how they have shaped our planet over billions of years.

Plate Tectonics: A Revolution in Earth Science

Now, let's delve into one of the most transformative theories in the field of geology—plate tectonics. This theory provides a comprehensive explanation for the movement of the Earth's lithospheric plates and the geological phenomena associated with their interactions.

Imagine the Earth's crust as a vast jigsaw puzzle with interlocking pieces. These pieces are the tectonic plates, and they are not fixed in place. Instead, they are in a constant state of motion, driven by forces beneath the Earth's surface.

The driving force behind plate tectonics is the convection of heat in the mantle. As heat rises from the Earth's interior, it causes the semi-solid rock in the mantle to become less dense and buoyant. This buoyant material then moves towards the Earth's surface, where it cools and becomes denser, subsequently sinking back into the depths. This continuous cycle of rising and sinking material creates convection cells in the mantle, which have the remarkable effect of dragging the tectonic plates along with them.

Plate Boundaries: Where Earth's Drama Unfolds

To fully appreciate the significance of plate tectonics, we must explore the various types of plate boundaries and the dynamic geological features they create.

Divergent Boundaries: At divergent boundaries, tectonic plates move away from each other. This movement leads to the stretching and thinning of the Earth's crust, resulting in the formation of rift zones and mid-ocean ridges. Mid-ocean ridges are underwater mountain ranges where new oceanic crust is continuously formed as magma rises from the mantle and solidifies. The Mid-Atlantic Ridge is a prominent example of a divergent boundary.

Convergent Boundaries: In contrast, convergent boundaries are where tectonic plates move toward each other. When an oceanic plate collides with a continental plate, the denser oceanic plate is forced beneath the continental plate in a process known as subduction. This subduction leads to the formation of deep ocean trenches and volcanic arcs on the Earth's surface. The Andes Mountains, along the western edge of South America, are a result of the subduction of the Nazca Plate beneath the South American Plate. When two continental plates collide, they can create massive mountain ranges, such as the Himalayas.

Transform Boundaries: Transform boundaries are characterized by horizontal movement, where tectonic plates slide past each other. The friction between these plates can create immense stress, which is eventually released in the form of earthquakes along fault lines. The San Andreas Fault in

California is perhaps the most famous transform boundary, where the Pacific Plate and the North American Plate are grinding against each other.

Understanding the various types of plate boundaries and their associated geological processes helps us grasp the dynamic nature of our planet's surface. It also provides insights into the distribution of earthquakes, volcanic activity, and the formation of mountain ranges.

The Fossils and Continental Drift Connection

The concept of plate tectonics was not always accepted in the scientific community. It took several decades of observations, data collection, and the development of supporting evidence for it to gain widespread acceptance. One key piece of evidence that played a pivotal role in validating the theory of plate tectonics was the distribution of fossils on different continents.

Paleontologists discovered that fossils of similar species were found on continents that are now widely separated by oceans. For example, fossils of the extinct reptile Mesosaurus were found in both South America and Africa. This distribution could not be explained unless these continents were once connected or in close proximity.

The hypothesis of continental drift, proposed by Alfred Wegener in the early 20th century, suggested that the continents were once part of a supercontinent called Pangaea. According to this idea, Pangaea began to break apart around 200 million years ago, eventually leading to the present-day positions of the continents. While Wegener's continental drift theory faced skepticism at the time, it laid the groundwork for the later acceptance of plate tectonics.

The key to understanding the movement of continents and the distribution of fossils lies in plate tectonics. The movement of tectonic plates explains how continents can drift apart and come together over geological time scales, providing a satisfying explanation for the distribution of fossils and the geological history of our planet.

The Power of Plate Tectonics: Earth's Dynamic Features

Plate tectonics is not just a theory; it's a dynamic process that has shaped our world in profound ways. Here are some of the remarkable geological features and phenomena associated with plate tectonics:

Mountain Building: The collision of continental plates at convergent boundaries is responsible for the creation of some of the world's tallest mountain ranges, including the Himalayas, the Rocky Mountains, and the Alps. The immense forces generated during these collisions cause the Earth's crust to fold, fault, and uplift, giving rise to majestic peaks.

Volcanoes and Island Arcs: At convergent boundaries, where an oceanic plate descends beneath another plate, the intense heat and pressure lead to the melting of rocks in the subducted plate. This molten rock, or magma, rises to the surface and can result in volcanic eruptions. Island arcs, such as the Japanese archipelago and the Aleutian Islands, are formed in this manner.

Earthquakes: The movement of tectonic plates along fault lines at transform boundaries generates immense stress. When this stress is released, it leads to earthquakes, which can range from minor tremors to devastating events. Earthquakes often occur along fault lines, which are fractures in the Earth's crust where rocks have moved relative to each other.

Ocean Basins: Divergent boundaries in the ocean create mid-ocean ridges, which are underwater mountain ranges. As tectonic plates move apart, magma rises from the mantle, solidifies, and forms new oceanic crust. Over millions of years, this process results in the expansion of ocean basins. The Atlantic Ocean, for example, is widening as the North American and Eurasian plates move away from the Eurasian and African plates.

Continental Drift: The motion of tectonic plates also leads to the shifting of continents. While the movement is relatively slow in human terms, it has significant geological implications.

Continents drift apart, collide, and sometimes join together over millions of years, reshaping Earth's landmasses.

Tsunamis: Subduction zones, where an oceanic plate is forced beneath a continental plate, can generate massive underwater earthquakes. These earthquakes have the potential to displace large volumes of seawater, leading to the formation of tsunamis, which are giant ocean waves that can devastate coastal areas when they reach the shore.

Understanding these geological features and phenomena is not only crucial for advancing our knowledge of the Earth but also for mitigating the risks associated with natural disasters. Earthquakes, volcanic eruptions, and tsunamis are all consequences of the dynamic nature of our planet's lithosphere, driven by the relentless movement of tectonic plates.

The Continuation of a Grand Idea

The theory of plate tectonics has revolutionized the field of earth science and our understanding of our planet. It has provided a unifying framework that explains a wide range of geological phenomena and has allowed us to interpret the Earth's history through the movements of continents and the formation of mountain ranges.

This grand idea has not only transformed our understanding of the Earth's past but also continues to shape our comprehension of the planet's future. Scientists study the movement of tectonic plates to predict earthquakes, volcanic eruptions, and the potential consequences of rising sea levels due to the shifting of continents.

Moreover, plate tectonics plays a crucial role in shaping the Earth's environment. The carbon cycle, for example, is intimately linked to the movement of tectonic plates. When oceanic plates are subducted into the mantle, they carry with them carbonate minerals. Through geological processes, these carbonates are eventually released as carbon dioxide (CO_2) during volcanic eruptions, contributing to the Earth's carbon cycle and climate regulation.

The Ongoing Journey of Exploration

As we conclude this exploration of geology and plate tectonics, it is essential to appreciate that our understanding of the Earth's inner workings is an ongoing journey. New discoveries, technological advancements, and interdisciplinary research continue to expand our knowledge of the dynamic processes that govern our planet.

From the depths of the Earth's mantle to the towering peaks of mountain ranges, from the rumblings of earthquakes to the eruption of volcanoes, the Earth's geology is a testament to the enduring power of curiosity and scientific inquiry. Our planet's geological history is written in its rocks, its mountains, and the very ground beneath our feet. By studying and embracing the wonders of our dynamic world, we gain insights into the forces that have shaped our past and continue to shape our future.

In the chapters that follow, we will further explore the Earth's climate and the water cycle, delving into the intricate interactions between geology and the environment. Together, we will unravel the intricate tapestry of our world and gain a deeper appreciation for the complex web of processes that govern how the world really works.

3.2: Earth's Climate and the Water Cycle

The Earth is not a static entity; it is a dynamic system in which every component is intricately connected. In this chapter, we venture into the world of Earth's climate and the water cycle—two fundamental aspects of our environment that are profoundly influenced by geology. As we explore these topics, we will gain a deeper understanding of how geological processes shape the Earth's climate and how the movement of water drives life on our planet.

The Dance of Earth's Climate: Geology's Influence

Earth's climate is a delicate balance that arises from complex interactions among the atmosphere, oceans, land, and life. While climate is often associated with atmospheric processes, the influence of geology is profound and far-reaching.

Continents and Climate: The presence and arrangement of continents play a crucial role in shaping Earth's climate. Landmasses absorb and release heat differently than oceans. For example, large landmasses like Asia can experience extreme temperature variations between summer and winter due to their distance from the moderating influence of the ocean. Mountains, another geological feature, can significantly impact climate by blocking or redirecting wind patterns, leading to rain shadows and differences in precipitation.

Volcanic Activity: Volcanic eruptions, often driven by geological processes at plate boundaries, can have a significant impact on global climate. When volcanoes release large amounts of ash and gases into the atmosphere, they can temporarily block sunlight and lower temperatures. The famous eruption of Mount Tambora in 1815, for example, led to the "Year Without a Summer" in 1816, resulting in crop failures and food shortages.

Tectonic Movements and Sea Level: Plate tectonics also

influence sea levels and ocean circulation, which in turn affect climate. Tectonic movements can alter the shape and size of ocean basins, influencing the distribution of ocean currents. Changes in sea level, driven by geological processes like the uplift of continental plates or the melting of ice sheets, can impact coastal climates and ecosystems.

The Water Cycle: Nature's Recycling System

Water is the lifeblood of our planet, and its movement through the water cycle sustains all living organisms. This cycle, often referred to as the hydrological cycle, involves the continuous circulation of water between the Earth's surface and the atmosphere. Geological processes are at the heart of this cycle.

Evaporation and Condensation: The water cycle begins with the evaporation of water from oceans, lakes, rivers, and even land surfaces. This process is driven by solar energy, which heats the Earth's surface and causes water to transform from liquid to vapor. As moist air rises, it cools and condenses into clouds, marking the beginning of precipitation.

Precipitation: Precipitation includes rain, snow, sleet, and hail—various forms of water returning to the Earth's surface. Geological features, such as mountains, can influence the distribution and amount of precipitation in different regions. Mountains can block moist air masses, leading to rain shadows on the leeward side, while promoting heavy precipitation on the windward side.

Runoff and Erosion: Once precipitation reaches the Earth's surface, it follows various pathways. Some of it flows directly into rivers and streams, becoming surface runoff. Geological factors like soil composition and slope determine the rate of runoff.

Additionally, the erosive power of water can shape landscapes over time, creating valleys, canyons, and river deltas.

Groundwater and Aquifers: Much of the water that infiltrates the ground becomes groundwater, stored in underground reservoirs called aquifers. Geological formations, including sedimentary rock layers, play a significant role in storing and transmitting groundwater. Aquifers are vital sources of freshwater for drinking, agriculture, and industry.

Return to the Oceans: Ultimately, water returns to the oceans through a variety of routes, including rivers, groundwater discharge, and direct precipitation over the ocean. This process completes the water cycle and ensures a continuous supply of freshwater for terrestrial and marine ecosystems.

The Role of Geology in Shaping Climate Zones

Climate zones, such as polar, temperate, and tropical, are defined by patterns of temperature and precipitation. Geology plays a pivotal role in creating and influencing these climate zones.

Tropical Rainforests and Equatorial Climates: Near the

equator, where sunlight is intense year-round, tropical rainforests thrive. These regions are characterized by lush vegetation and high levels of rainfall. The proximity of equatorial regions to the oceans, combined with low-lying coastal plains, contributes to their specific climate.

Deserts and Rain Shadows: On the other side of the spectrum, deserts often form on the leeward side of mountains in rain shadow areas. As moist air rises over mountains and cools, it releases moisture as rain on the windward side. By the time the air descends on the leeward side, it has lost most of its moisture, creating arid conditions. The Sahara Desert and the Atacama Desert are examples of such arid regions influenced by geological features.

Polar Climates and Glacial Landscapes: Polar climates, found near the Earth's poles, are profoundly affected by the distribution of land and ice. Geological processes related to tectonics and the shifting of continents have led to the positioning of Antarctica over the South Pole. The vast ice sheets covering Antarctica and Greenland reflect sunlight, maintain cold temperatures, and influence global climate patterns.

Temperate Zones and Continental Climate: Temperate zones, which include many of the world's most populous regions, experience distinct seasons. The moderating influence of oceans helps create a more temperate climate near coastlines, while continental interiors often exhibit greater temperature extremes due to their distance from the ocean.

Geology's Legacy: Fossils and Climate Records

Geology provides us with a valuable record of past climates through the study of fossils and sedimentary rocks. Fossils, the preserved remains of ancient life forms, offer insights into the types of organisms that once inhabited a region and the environmental conditions of the time. For example, the presence of fossilized marine creatures in sedimentary rocks found in deserts indicates that these areas were once underwater.

Sedimentary rocks, which form through the accumulation and cementation of sediments, can also hold clues about past climates. Layers of sedimentary rock can reveal shifts in sea level, ancient river systems, and even evidence of past glaciations. By analyzing these geological records, scientists can reconstruct the Earth's climate history, helping us understand the natural climate variability that has occurred over geological time scales.

Climate Change and Geology: A Two-Way Interaction

In recent years, the topic of climate change has gained widespread attention. While climate change discussions often focus on human activities and greenhouse gas emissions, geology plays a crucial role in understanding this complex issue.

The Carbon Cycle: The movement of carbon through the Earth's atmosphere, oceans, and geological reservoirs is known as the carbon cycle. Geological processes, such as the weathering of rocks and the burial of organic matter, play a pivotal role in regulating carbon dioxide (CO2) levels in the atmosphere. Volcanic activity, as a geological process, releases CO2 into the atmosphere. Understanding the carbon cycle is essential for comprehending the drivers of both natural and human-induced climate change.

Climate Archives in Ice and Sediments: Geological archives in ice cores and sediment layers provide critical data for understanding past climate variations. Ice cores from polar regions contain layers that represent different periods in Earth's history, offering insights into temperature, atmospheric composition, and past climate events. Sediments in oceans and lakes preserve a record of ancient climates through the deposition of materials like pollen, minerals, and microorganisms.

Geological Hazards and Climate Adaptation: Geology also plays a vital role in assessing climate-related hazards. Rising sea levels, for example, can increase the vulnerability of coastal regions to storm surges and flooding. Understanding the geological characteristics of coastal areas is essential for effective climate adaptation and disaster preparedness.

Embracing the Complexity of Earth's Systems

As we navigate the intricacies of Earth's climate and the water cycle, it becomes clear that our planet's functioning is a complex and interconnected web of processes. Geology is an integral part of this tapestry, influencing climate patterns, shaping landscapes, and preserving the history of our world.

This section has provided a glimpse into how geological processes impact climate zones, the water cycle, and climate change. Yet, it is only the beginning of our exploration into the intricate relationships that govern how the world really works. In the chapters ahead, we will continue our journey of discovery, delving deeper into the Earth's dynamic biosphere, the evolution of life, and the mysteries of our natural world. Each step brings us closer to embracing the wonders of our planet and understanding the true nature of our interconnected existence.

3.3: The Dynamic Biosphere

The Earth is a living planet, teeming with a stunning array of life forms that have adapted to a wide range of environments. From the deepest oceans to the highest mountaintops, life has found a way to thrive. In this chapter, we delve into the dynamic biosphere, the living realm of our planet, and explore how geology plays a fundamental role in shaping the Earth's diverse ecosystems, from the structure of landscapes to the composition of soils.

The Living Earth: A Symphony of Biodiversity

The biosphere is the realm of life on Earth, encompassing all living organisms, from microscopic bacteria to towering trees and from the depths of the Earth's crust to the upper reaches of the atmosphere. It is a complex web of interactions and interdependencies, where each species plays a unique role in maintaining the delicate balance of life.

Biodiversity is a hallmark of the biosphere, reflecting the incredible variety of life forms and ecosystems on our planet. It encompasses the diversity of species, genetic diversity within those species, and the diversity of ecosystems in which they reside. Biodiversity is not just a source of wonder; it also underpins the ecological services that sustain human societies, from providing clean air and water to supporting agriculture and medicine.

Geology and the Formation of Habitats

Geological processes play a pivotal role in creating and shaping the habitats that support life on Earth. From the formation of mountains to the carving of valleys, geology influences the physical structure of landscapes and the distribution of habitats. Here are some key geological factors that impact the biosphere:

Mountain Building: The collision of tectonic plates and the uplift of continental crust give rise to mountain ranges. Mountains create a range of habitats, from high alpine meadows to rugged slopes. They also influence weather patterns by altering wind and moisture patterns, leading to differences in temperature and precipitation on windward and leeward sides, a phenomenon known as orographic effect.

Valleys and Floodplains: River valleys and floodplains are created and shaped by geological processes like erosion and sediment deposition. These fertile areas support diverse ecosystems and are often important for agriculture. Flooding, a natural geological process, replenishes nutrients in these habitats.

Caves and Karst Landscapes: Karst landscapes, characterized by limestone formations, caves, and sinkholes, are shaped by the dissolution of soluble rocks. These underground environments provide unique habitats for specialized organisms, including troglobites, species adapted to life in caves.

Coastal Zones: Coastal ecosystems are profoundly influenced by geological features such as shorelines, cliffs, and barrier islands. Coastal geology affects the dynamics of tides, currents, and erosion, influencing the distribution of marine life and the resilience of coastal communities.

Geology and Soil Formation

Soil is a vital component of the biosphere, serving as the foundation for terrestrial ecosystems. Geology plays a significant role in soil formation and composition. Soil is more than just dirt; it is a complex mixture of minerals, organic matter, water, air, and countless microorganisms. The parent material from which soils develop is primarily influenced by underlying geological formations.

Parent Material: The geological parent material, or bedrock, determines the mineral content of soils. Granite bedrock, for example, yields soil rich in minerals like quartz, feldspar, and mica, while limestone bedrock contributes calcium-rich soils. These mineral compositions influence the availability of nutrients to plants and microorganisms.

Weathering: Weathering, a geological process, breaks down rocks into smaller particles over time. Mechanical weathering, such as freeze-thaw cycles or the action of plant roots, physically breaks rocks apart. Chemical weathering involves the alteration of rock minerals through chemical reactions. Both types of weathering contribute to soil formation.

Soil Horizons: Soils are typically organized into layers, or horizons, each with distinct characteristics. The O horizon, also known as the organic horizon, contains decomposing organic matter like leaves and plant debris. Below it, the A horizon, or topsoil, is rich in organic material and microorganisms, making it fertile. The B horizon, or subsoil, accumulates minerals leached from above. The C horizon consists of weathered parent material, and the R horizon is unweathered bedrock.

Soil Profiles: The unique combination of soil horizons and parent material gives rise to specific soil profiles, each with its properties and suitability for different types of vegetation. Soil profiles influence the types of plants that can grow in an area and, consequently, the composition of ecosystems.

The Water-Earth Connection: Aquatic Ecosystems

While we often associate geology with terrestrial landscapes, it also plays a vital role in shaping aquatic ecosystems. Oceans, rivers, lakes, and wetlands are profoundly influenced by geological processes, from the composition of sediments on the ocean floor to the formation of river valleys.

Ocean Basins: Geological features on the ocean floor, such as seamounts, trenches, and mid-ocean ridges, influence ocean circulation, nutrient availability, and the distribution of marine life. For example, mid-ocean ridges are hotspots of biodiversity due to the upwelling of nutrient-rich waters.

River Systems: Rivers and their watersheds are shaped by geological processes, including the erosion of landscapes and the deposition of sediments. River ecosystems are highly dynamic, with habitats ranging from fast-flowing streams to slow-moving meanders. The geology of riverbeds affects water quality and habitat availability for aquatic organisms.

Lakes and Wetlands: Geological factors, such as glacial activity or tectonic subsidence, can create depressions that fill with water, forming lakes and wetlands. These ecosystems are characterized by unique hydrological and geological features, and they often support specialized plant and animal species.

Aquifers and Groundwater: Underground aquifers, geological formations that store water, are essential for supplying freshwater to terrestrial and aquatic ecosystems. They can sustain wetlands, springs, and rivers, even during dry periods, contributing to biodiversity and ecosystem stability.

Geological Time and the Evolution of Life

The geological time scale spans billions of years, providing the backdrop for the evolution of life on Earth. Fossils, preserved remains of ancient organisms, are geological archives that offer insights into the history of life. Geological processes, such as sedimentation, preservation, and fossilization, shape the fossil record.

Sedimentation: Over geological time, sediments accumulate in various environments, including oceans, lakes, and riverbeds. These sediments can entomb the remains of plants and animals, preserving them as fossils. Fossil-rich sedimentary rock layers provide a window into past ecosystems.

Preservation: Fossils are often preserved when the remains are buried rapidly, preventing decomposition or scavenging.

Certain geological conditions, such as the presence of minerals or low oxygen levels, can aid in fossil preservation.

Fossilization: The process of fossilization involves the gradual replacement of organic material with minerals. Over time, minerals seep into the microscopic spaces within an organism's remains, creating a mineralized replica. Fossilization can occur in various ways, including permineralization, petrification, and carbonization.

Dating Fossils: Geologists use various dating techniques, including radiometric dating, to determine the age of fossils and the rocks in which they are found. This allows us to reconstruct the timeline of life's evolution and understand the changing composition of Earth's ecosystems over time.

Geological Hazards and Ecosystem Resilience

While geological processes are essential for shaping habitats and preserving fossils, they can also pose significant challenges to ecosystems and human communities. Geological hazards, such as earthquakes, volcanic eruptions, landslides, and tsunamis, have the potential to disrupt ecosystems, destroy habitats, and alter landscapes.

Ecological Succession: Ecosystems have developed strategies to cope with disturbances like volcanic eruptions and landslides. Ecological succession is the process by which ecosystems recover and rebuild after a disturbance. Pioneer species, often adapted to harsh conditions, are the first to colonize disturbed areas. Over time, these species pave the way for more complex communities to establish themselves.

Biodiversity and Resilience: Biodiversity is a key factor in the resilience of ecosystems to geological hazards. Diverse ecosystems tend to recover more quickly from disturbances and may be better equipped to adapt to changing conditions. Protecting biodiversity is not only crucial for preserving Earth's rich tapestry of life but also for enhancing ecosystem resilience in the face of natural challenges.

The Future of the Dynamic Biosphere

As we continue to explore the intricate relationship between geology and the biosphere, it becomes clear that our understanding of Earth's living systems is an ongoing journey. The dynamic biosphere is a testament to life's adaptability and resilience, shaped by geological processes over billions of years.

In the chapters that follow, we will delve deeper into the evolution of life, the genetic mechanisms that drive biodiversity, and the complex interactions within ecosystems. Each step brings us closer to embracing the wonders of our planet's living realm and understanding the true nature of our interconnected existence.

Chapter 4: Evolution & Biodiversity

4.1: The Theory of Evolution

In the grand tapestry of life on Earth, few scientific theories have had as profound an impact as the theory of evolution. It is a unifying concept that explains the diversity of life, the relationships between species, and the processes by which organisms adapt and change over time. This chapter embarks on a journey through the theory of evolution, exploring its historical roots, its core principles, and its far-reaching implications for our understanding of the living world.

The Historical Context: Before the Theory of Evolution

Before delving into the theory of evolution itself, it's essential to understand the historical context in which it emerged. For much of human history, the diversity of life on Earth was explained through creation myths and religious narratives. These accounts attributed the existence of species to the creative acts of gods or other supernatural entities.

The idea that species could change over time and that life had a natural, rather than supernatural, origin was a radical departure from prevailing beliefs. The foundations of the theory of evolution began to take shape in the 18th and 19th centuries, laying the groundwork for a revolutionary paradigm shift in our understanding of life.

Precursors to Evolutionary Thinking

Several thinkers and naturalists made important contributions that paved the way for the theory of evolution. These precursors to evolutionary thinking challenged traditional views of life's immutability and the fixity of species:

Buffon and the Idea of Environmental Influence: Georges-Louis Leclerc, Comte de Buffon, proposed that environmental factors could gradually change species over time. He suggested that animals could adapt to their environments through a process of transformation. While Buffon's ideas were not a full-fledged theory of evolution, they hinted at the possibility of species change.

Lamarck and the Inheritance of Acquired Traits: Jean-Baptiste Lamarck proposed the idea that organisms could acquire new traits during their lifetimes and pass them on to their offspring. He believed that the use or disuse of specific organs could lead to their development or degeneration. Although Lamarck's theory was later discredited, it introduced the concept of species change through inheritance of acquired characteristics.

Cuvier and the Concept of Catastrophism: Georges Cuvier, a pioneering paleontologist, introduced the concept of catastrophism, which suggested that the Earth had been shaped by a series of catastrophic events followed by the creation of new species. While Cuvier did not propose a mechanism for species change, his work on extinction challenged the notion of species' unchanging nature.

Darwin and Wallace: The Theory of Natural Selection

The theory of evolution as we know it today emerged most prominently through the work of Charles Darwin and Alfred Russel

Wallace. Their independent discoveries of the same fundamental principle—natural selection—transformed our understanding of life's diversity and adaptation.

Charles Darwin: Darwin's journey on the HMS Beagle in the early 1830s was a pivotal moment in the development of evolutionary theory. During his voyage, he collected a vast array of specimens and observed the remarkable diversity of life on Earth. He also encountered fossils and geological features that hinted at an ancient Earth with a complex history.

Darwin's key insight came from his observations of variation within populations and the realization that some traits provided a selective advantage in specific environments. He reasoned that, over time, those individuals with advantageous traits would be more likely to survive, reproduce, and pass on those traits to their offspring. This process, which he termed "natural selection," became the central mechanism of his theory of evolution.

In 1859, Darwin published "On the Origin of Species," in which he presented his theory of evolution by natural selection. He argued that species could change over time through this process, with new species arising from common ancestors. His book revolutionized biology and ignited a spirited debate that continues to this day.

Alfred Russel Wallace: While Darwin's work is synonymous with the theory of evolution, Alfred Russel Wallace independently developed a similar theory of natural selection. In 1858, Wallace sent a letter to Darwin outlining his ideas, which closely mirrored Darwin's own thoughts on the subject.

Recognizing the importance of Wallace's ideas, Darwin and

Wallace jointly presented their findings to the Linnean Society of London in 1858. This event marked the beginning of a fruitful collaboration between the two naturalists. Darwin's more extensive body of work and publications led to greater recognition, but both men made significant contributions to the theory of evolution.

The Core Principles of the Theory of Evolution

The theory of evolution by natural selection is built upon several foundational principles that provide a framework for understanding how species change and diversify over time:

Variation: Within any population of organisms, there is genetic variation. Individuals within a species are not identical; they have different traits and characteristics. This variation arises from mutations, genetic recombination, and other processes.

Heritability: Traits that vary among individuals are often passed on from one generation to the next. Offspring tend to inherit the traits of their parents, although variations can occur due to genetic mutations.

Overproduction: Most species produce more offspring than can survive to adulthood. This leads to competition for limited resources among the offspring, which creates a "struggle for existence."

Differential Survival and Reproduction: In the struggle for existence, individuals with traits that give them a survival or reproductive advantage are more likely to thrive and pass on their advantageous traits to the next generation. This process is the essence of natural selection.

Adaptation: Over time, natural selection results in the accumulation of advantageous traits within a population, leading to adaptations that enhance an organism's fitness in its specific environment. Adaptations are features or behaviors that increase an organism's chances of survival and reproduction.

The Evidence for Evolution

The theory of evolution is supported by a wealth of evidence from various fields of science, including paleontology, comparative anatomy, embryology, biogeography, and genetics. Each of these lines of evidence provides insights into the patterns and mechanisms of evolution:

Fossil Record: Fossils are the preserved remains or traces of ancient organisms. The fossil record reveals a succession of life forms through time, with simpler forms appearing in older rock layers and more complex forms in younger layers. Transitional fossils, which display characteristics of both older and more recent species, provide evidence for gradual changes over time.

Comparative Anatomy: Comparative anatomy examines the structural similarities and differences among species. Homologous structures, such as the forelimbs of vertebrates, share a common evolutionary origin, even if they serve different functions in different species. Vestigial structures, which have no apparent function in an organism, are remnants of ancestral traits.

Embryology: The study of embryonic development reveals similarities in the early stages of development among different species. This shared embryonic heritage is indicative of common ancestry.

Biogeography: Biogeography examines the distribution of species across different geographic regions. Patterns of distribution often align with the historical movement of continents and the isolation of populations, reflecting the influence of both geological and evolutionary processes.

Molecular Genetics: Advances in molecular biology and genetics have provided powerful evidence for evolution. DNA sequencing allows scientists to compare the genetic makeup of different species, revealing genetic relationships and tracing common ancestry. The genetic code itself is shared among all living organisms, providing a unifying link in the tree of life.

The Tree of Life: Unity and Diversity

The theory of evolution is encapsulated in the metaphorical "tree of life," a visual representation of the genealogical relationships among all living organisms. This tree illustrates the concept that all species share a common ancestry and have evolved from a single common ancestor.

At the base of the tree of life are the earliest life forms, such as single-celled organisms. As one ascends the tree, branches represent the divergence of species over time, with more recent branches representing organisms that share a more recent common ancestor. The branching structure of the tree of life embodies the principle of descent with modification, a core tenet of evolutionary theory.

Speciation: The Origin of New Species

One of the central questions in evolutionary biology is how new species arise. Speciation is the process by which a single population splits into two or more distinct species. It occurs when populations become reproductively isolated, meaning they can no longer interbreed and produce fertile offspring.

Speciation can occur through several mechanisms:

Allopatric Speciation: This form of speciation occurs when a population becomes geographically isolated from the rest of its species. Over time, genetic differences accumulate between the isolated population and the parent population, leading to the formation of new species.

Sympatric Speciation: Sympatric speciation occurs when populations within the same geographic area become reproductively isolated due to factors other than geographic barriers. This can happen through mechanisms such as polyploidy (an increase in the number of sets of chromosomes) or ecological specialization.

Parapatric Speciation: In parapatric speciation,

populations that are adjacent to each other but have limited gene flow can diverge and form new species. This can occur when there is a gradient of environmental conditions that create selection pressures.

The Impact of Evolutionary Theory

The theory of evolution has far-reaching implications for our understanding of the natural world and its impact on diverse fields:

Biology and Medicine: Evolutionary biology is essential for understanding disease resistance, antibiotic resistance, and the emergence of new diseases. It also provides insights into the development of new drugs and medical treatments.

Ecology: Understanding evolutionary processes is crucial for studying the dynamics of ecosystems, including predator-prey relationships, competition, and the coevolution of species.

Agriculture: Evolutionary principles are used in crop breeding to develop more resilient and productive plant varieties. They also inform pest management strategies.

Conservation: Evolutionary theory plays a key role in conservation biology, helping to preserve endangered species and restore ecosystems.

Paleontology: Evolutionary concepts guide the interpretation of the fossil record, shedding light on the history of life on Earth.

Anthropology: Evolutionary theory is fundamental to the study of human origins and the development of cultural and behavioral traits in our species.

The Continuing Evolution of Evolutionary Theory

While Charles Darwin and Alfred Russel Wallace laid the foundation for evolutionary theory, our understanding of evolution has continued to evolve itself. Modern advances in genetics, genomics, and molecular biology have provided new insights into the mechanisms of evolution.

Contemporary evolutionary biology explores topics such as genetic drift, gene flow, sexual selection, and coevolution in greater depth. Researchers also study the role of symbiosis and horizontal gene transfer in shaping the genomes of organisms.

Additionally, the field of evolutionary developmental biology, or "evo-devo," examines the genetic and developmental processes that underlie evolutionary change. This interdisciplinary approach bridges genetics, developmental biology, and evolutionary biology to elucidate how changes in the genetic toolkit of organisms lead to morphological diversity.

The Beauty of a Unifying Theory

The theory of evolution by natural selection is a remarkable scientific achievement that has transcended its origins to become a unifying principle in biology. It ties together the diversity of life on Earth, explains the adaptive features of organisms, and provides a coherent framework for understanding the living world.

In embracing the theory of evolution, we gain insight into the interconnectedness of all life forms and our place within the tree of life. It is a testament to the power of scientific inquiry and human curiosity—a journey that continues to reveal the mysteries of our world and the rich tapestry of life that surrounds us.

4.2: The Tree of Life: Biodiversity and Ecosystems

Beneath the lush canopies of rainforests, beneath the waves of vast oceans, and within the most inhospitable deserts, Earth's ecosystems pulsate with life. The intricate dance of organisms, the tapestry of species, and the delicate balance of nature have been sculpted by billions of years of evolution. In this chapter, we venture into the heart of biodiversity and ecosystems, seeking to unravel the beauty and complexity of the natural world.

Biodiversity: The Symphony of Life

Biodiversity is a term that encapsulates the dazzling variety of life on Earth. It encompasses not just the number of species but also the genetic diversity within those species and the diverse ecosystems that house them. Biodiversity is the result of an ongoing evolutionary process that has given rise to an astonishing array of life forms.

Species Diversity: At the core of biodiversity lies the diversity of species that inhabit our planet. While scientists have cataloged and described roughly 2 million species, it is estimated that millions more remain undiscovered. The catalog includes microorganisms, fungi, plants, animals, and more, each with its unique characteristics and role in the ecological tapestry.

Genetic Diversity: Genetic diversity, found within each species, represents the variability in the genetic makeup of individuals. It arises from mutations, genetic recombination, and other genetic processes. This diversity serves as the raw material for evolution, enabling species to adapt to changing environments.

Ecosystem Diversity: The Earth's surface is a patchwork of ecosystems, each with its unique set of organisms and environmental conditions. These ecosystems range from the towering forests of the Pacific Northwest to the stark deserts of the Sahara, and from the bustling coral reefs of the tropics to the frigid tundras of the polar regions. Ecosystem diversity is crucial as each ecosystem provides essential services and habitats.

The Tree of Life: A Genealogical Odyssey

The concept of the tree of life serves as a powerful visual metaphor for the genealogical relationships among all living organisms on Earth. It illustrates the interconnectedness of life, with every species representing a branch on this sprawling arboreal structure.

A Common Ancestry: Central to the tree of life is the idea that all life on Earth shares a common ancestry—an ancient, single-celled organism from which all subsequent life forms descended. This concept underscores the unity of life and its interconnectedness.

Branches and Divergence: As life unfolded, lineages branched and diverged, giving rise to new species. These branching points on the tree represent speciation events, where one lineage splits into two or more distinct lineages. The process of descent with modification is the bedrock of evolutionary theory.

Taxonomy: To make sense of this vast array of life, scientists employ taxonomy—a hierarchical system for classifying and naming organisms. Taxonomy begins with broad domains (e.g., Bacteria, Archaea, Eukarya) and proceeds to more specific ranks, such as kingdoms, phyla, classes, orders, families, genera, and species.

Phylogenetics: Phylogenetics, the study of evolutionary relationships, helps construct the branches of the tree of life. By analyzing genetic and morphological data, scientists can trace the branching patterns and understand how different species are related.

Ecosystems: The Dynamic Communities

Ecosystems are the living engines that power the biosphere. They encompass intricate communities of organisms interacting with each other and their physical environments. Ecosystems are as diverse as the species they house, each one finely tuned to its specific set of conditions.

Biotic and Abiotic Elements: Ecosystems are a tapestry woven from both biotic (living) and abiotic (non-living) threads. The biotic components include organisms like plants, animals, and microorganisms, while abiotic factors encompass climate, soil, water, and geological features.

Energy Flow: The heartbeat of ecosystems is the flow of energy. Sunlight is captured by photosynthesis in plants, converting it into chemical energy. This energy cascades through the food web as organisms consume one another. Ultimately, energy dissipates as heat, maintaining the cycle of life.

Nutrient Cycling: Nutrient cycling is the ecological recycling system. Elements like carbon, nitrogen, and phosphorus are essential for life, and they move through ecosystems in a continuous loop. Decomposers, such as bacteria and fungi, play a pivotal role in breaking down organic matter and returning nutrients to the ecosystem.

Biodiversity and Stability: Biodiversity within ecosystems contributes to their stability and resilience. Diverse ecosystems often exhibit higher productivity, greater resistance to disturbances, and better recovery from disruptions. The various species within an ecosystem have distinct roles, or niches, that help maintain balance.

Ecological Interactions: The Choreography of Nature

Within the intricate dance of ecosystems, organisms engage in a multitude of interactions that define the dynamics of populations and communities. These interactions shape the web of life and underscore the interdependence of species.

Predation: Predation, a fundamental interaction, involves one organism—the predator—hunting and consuming another—the prey. Predation has far-reaching effects on the abundance and distribution of species within ecosystems and drives adaptations in both predators and prey.

Competition: Competition occurs when different species or individuals within a species vie for limited resources like food, water, or shelter. Intense competition can lead to niche differentiation, where species evolve to occupy distinct ecological niches to reduce competition.

Symbiosis: Symbiosis represents close and long-term interactions between two different species. It encompasses mutualism, where both species benefit; commensalism, where one benefits while the other is unaffected; and parasitism, where one benefits at the expense of the other.

Pollination: Pollination is a prime example of mutualism in action. Flowering plants and their pollinators, including bees, butterflies, and birds, engage in a mutually beneficial relationship. Pollinators transfer pollen between flowers, facilitating plant reproduction and the production of fruits and seeds.

Human Impact on Biodiversity and Ecosystems: A Disruptive Tune

Despite the resilience and adaptability of Earth's ecosystems, they now confront unprecedented challenges driven by human activities. These threats range from habitat destruction and pollution to climate change and the introduction of invasive species, jeopardizing the stability and resilience of ecosystems and the survival of many species.

Habitat Destruction: The transformation of natural habitats into agricultural land, urban areas, and industrial zones leads to habitat loss and fragmentation. This diminishes the availability of suitable habitats for many species, making it increasingly challenging for them to survive.

Pollution: Pollution, which includes air, water, and soil pollution, can harm both terrestrial and aquatic ecosystems. Chemical pollutants, such as pesticides and industrial chemicals, can have toxic effects on organisms and disrupt food chains.

Overexploitation: Overharvesting of resources, such as overfishing and deforestation, can deplete populations of species and damage ecosystems. Unsustainable practices can lead to the collapse of fisheries and the degradation of forests.

Invasive Species: The introduction of non-native

species to new environments can disrupt ecosystems and outcompete native species. Invasive species can alter the composition and functioning of ecosystems, leading to biodiversity loss.

Climate Change: Climate change, primarily driven by the release of greenhouse gases, is altering the distribution of species and their habitats. Rising temperatures, shifting precipitation patterns, and ocean acidification pose significant challenges to biodiversity.

Conservation and Restoration: Guardians of the Tree of Life

Efforts to conserve biodiversity and restore ecosystems are crucial to mitigating the human impact on the natural world. Conservation actions encompass a range of strategies, from establishing protected areas and wildlife corridors to breeding programs for endangered species and habitat restoration projects.

Protected Areas: Protected areas, such as national parks and reserves, serve as sanctuaries for species and habitats. They help preserve biodiversity and provide opportunities for research on natural ecosystems.

Wildlife Corridors: Wildlife corridors link fragmented

habitats, enabling species to move and exchange genes. These corridors are vital for maintaining genetic diversity within populations.

Species Conservation: Conservation programs focus on the protection and recovery of endangered and threatened species. These efforts may involve breeding programs, habitat restoration, and reintroduction into the wild.

Habitat Restoration: Habitat restoration seeks to revitalize degraded ecosystems by removing invasive species, replanting native vegetation, and improving the physical and chemical conditions of habitats.

Sustainable Practices: Promoting sustainable practices in agriculture, forestry, fisheries, and other industries can reduce the negative impact on ecosystems and biodiversity.

The Ethical Imperative: Guardians of the Earth

The conservation of biodiversity and the protection of ecosystems extend beyond scientific concern—they represent ethical imperatives. Biodiversity embodies the culmination of billions of years of evolutionary history, and each species contributes to the beauty and complexity of life on Earth.

Intrinsic Value: Each species has intrinsic value, meaning it has worth in and of itself, irrespective of its utility to humans. Every species has a right to exist and flourish.

Interconnectedness: Biodiversity sustains the stability and functioning of ecosystems. Interconnected species support one another and contribute to the resilience of natural systems.

Cultural and Aesthetic Value: Biodiversity is an integral part of human culture, art, and spirituality. It has inspired countless cultures throughout history and continues to do so today.

Embracing the Web of Life: Our Journey Continues

In understanding the intricate dance of biodiversity, the sprawling tree of life, and the dynamic ecosystems that cradle them, we glimpse the profound interdependence of all living things. The web of life stands as a testament to the tenacity and adaptability of Earth's biosphere, shaped by billions of years of evolution.

As we embark on the journey to comprehend and safeguard the tree of life, we are called to stewardship, conservation, and an ethical commitment to preserve the diversity of life on our planet. It is a journey that enriches our scientific knowledge, deepens our appreciation for the natural world, and reminds us of our role as guardians of Earth's magnificent tapestry.

4.3: Genetics and Heredity: The Code of Life

Within the double-helix structure of DNA lies the code of life itself. Genetics, the study of heredity and the variation of inherited traits, unlocks the secrets of our biological existence. This chapter delves into the fascinating world of genetics, from the discovery of DNA to the intricacies of inheritance patterns, and explores the transformative power of genetic engineering.

The Birth of Genetics: From Mendel to DNA

The study of genetics has a rich history, marked by milestones that revolutionized our understanding of heredity and the transmission of traits from one generation to the next.

Mendel and the Laws of Inheritance: In the mid-19th century, Gregor Mendel, an Austrian monk, conducted pioneering experiments with pea plants. He discovered the fundamental principles of inheritance, known as Mendelian genetics. Mendel's laws—of segregation and independent assortment—revealed that genes come in pairs and are passed from parents to offspring in predictable ways.

Chromosomes and Heredity: In the early 20th century, scientists began to link Mendel's laws to the behavior of chromosomes during cell division. Chromosomes were found to carry genes, and their segregation and assortment explained the patterns of inheritance.

Discovery of the DNA Double Helix: In 1953, James Watson and Francis Crick unveiled the structure of DNA—a double helix made up of nucleotide building blocks. This discovery marked a turning point in genetics, as it revealed the molecular basis of heredity.

The DNA Molecule: The Blueprint of Life

DNA (deoxyribonucleic acid) is a remarkable molecule that encodes the genetic instructions for building and maintaining an organism. It is composed of four types of nucleotide bases: adenine (A), thymine (T), cytosine (C), and guanine (G).

These bases pair up in a specific manner (A with T and C with G), forming the rungs of the DNA ladder.

The Double Helix: The DNA molecule consists of two long chains of nucleotides running in opposite directions, coiled around each other to form a double helix. The complementary base pairing ensures that each strand can serve as a template for the synthesis of a new strand during replication.

Genes and Genomes: Genes are specific sequences of DNA that code for proteins or functional RNA molecules. The complete set of an organism's genes is known as its genome. Genes determine an organism's traits, from eye color to susceptibility to certain diseases.

Replication and Cell Division: DNA must be faithfully replicated before cell division so that each new cell receives an identical copy of the genetic information. The enzyme DNA polymerase catalyzes the synthesis of new DNA strands, ensuring the accuracy of replication.

Mutations: Mutations are changes in the DNA sequence. They can occur spontaneously or as a result of external factors like radiation or chemicals. While some mutations are harmful, others can provide the raw material for evolution by introducing genetic diversity.

Inheritance Patterns: Unraveling the Genetic Code

The transmission of traits from one generation to the next follows specific patterns, which have been studied and classified by geneticists.

Mendelian Inheritance: Mendelian inheritance, based on Mendel's laws, describes the inheritance of single gene traits. It includes dominant and recessive alleles, where dominant traits mask the expression of recessive traits in heterozygous individuals.

Incomplete Dominance: Incomplete dominance occurs when neither allele is completely dominant, resulting in an intermediate phenotype in heterozygous individuals. For example, in snapdragons, red (RR) and white (WW) alleles produce pink (RW) flowers.

Codominance: Codominance is a pattern where both alleles in a heterozygous individual are fully expressed. This is seen in traits like blood type, where individuals with one A allele and one B allele have type AB blood.

Polygenic Inheritance: Many traits are controlled by multiple genes, a phenomenon known as polygenic inheritance. Traits like height, skin color, and intelligence are influenced by the combined effects of many genes.

Sex-Linked Inheritance: Some genes are located on the sex chromosomes (X and Y). Genes on the X chromosome can be inherited differently by males and females, leading to sex-linked traits like color blindness and hemophilia.

Genetic Disorders: From Mutation to Disease

Genetic disorders result from mutations in specific genes, leading to abnormal or non-functional proteins and, consequently, health problems. These disorders can be inherited from parents or arise spontaneously.

Single Gene Disorders: Single gene disorders, also known as Mendelian disorders, are caused by mutations in a single gene. Examples include cystic fibrosis, sickle cell anemia, and Huntington's disease.

Chromosomal Disorders: Chromosomal disorders involve abnormalities in the structure or number of chromosomes. Down syndrome, for instance, is caused by the presence of an extra copy of chromosome 21.

Complex Disorders: Many common disorders, such as diabetes, heart disease, and cancer, have complex genetic components. They result from interactions between multiple genes and environmental factors.

Genetic Engineering: Shaping the Future of Life

The ability to manipulate genes has given rise to the field of genetic engineering, which offers incredible opportunities and raises ethical questions about the modification of living organisms.

Recombinant DNA Technology: Recombinant DNA technology allows scientists to combine DNA from different sources to create genetically modified organisms (GMOs). This technology has applications in medicine, agriculture, and biotechnology.

Gene Therapy: Gene therapy aims to treat genetic disorders by introducing or repairing genes in a patient's cells. It has the potential to cure previously untreatable diseases, although challenges remain in its development and implementation.

CRISPR-Cas9: The CRISPR-Cas9 system is a revolutionary gene-editing tool that enables precise modification of genes. It has opened new avenues for genetic research and therapeutic applications, but ethical considerations surround its use.

Genetically Modified Crops: Genetically modified (GM) crops have been engineered to resist pests, tolerate herbicides, and enhance nutritional content. They have the potential to address food security and reduce the environmental impact of agriculture.

Ethical and Social Implications: The ethical use of genetic engineering raises questions about consent, safety, and potential misuse. Society must grapple with the ethical and social implications of genetic manipulation.

The Future of Genetics: Unlocking the Code of Life

Genetics continues to evolve rapidly, with ongoing research unlocking new insights into the genome and its functions. Emerging fields like epigenetics explore how environmental factors influence gene expression and inheritance.

Epigenetics: Epigenetics investigates changes in gene expression that do not involve alterations to the DNA sequence. These changes can be influenced by environmental factors and can be passed on to future generations.

Personalized Medicine: Advances in genetics are paving the way for personalized medicine, where treatments and medications are tailored to an individual's genetic makeup. This approach promises more effective and targeted healthcare.

Genomic Medicine: Genomic medicine utilizes the information in an individual's genome to diagnose, treat, and prevent disease. It has the potential to revolutionize healthcare by providing precise and personalized interventions.

Genetic Research and Conservation: Genetics plays a crucial role in understanding and conserving biodiversity. Genetic research informs conservation efforts, helping to protect endangered species and preserve genetic diversity.

Conclusion: Decoding the Genetic Symphony

The study of genetics and heredity is an exploration of the code of life that defines us. From Mendel's peas to the double helix of DNA and the revolutionary CRISPR-Cas9, genetics has transformed our understanding of biology and medicine.

As we continue to unravel the mysteries of genetics, we must navigate ethical and societal questions about the boundaries of genetic engineering and the implications of our newfound powers. Genetics offers the promise of cures for genetic disorders, improved agricultural practices, and personalized healthcare, but it also demands responsible stewardship of the genetic code that shapes our existence.

In the ongoing symphony of life, genetics is the conductor, orchestrating the intricate movements of genes, traits, and diversity. It is a journey of discovery that unlocks the potential to shape the future of life on Earth.

Chapter 5: Genetics and Heredity

5.1 DNA and Genes
The Blueprint of Life

DNA, or deoxyribonucleic acid, is the fundamental building block of life, a complex molecule that carries the genetic instructions used in the growth, development, functioning, and reproduction of all known living organisms and many viruses. Resembling a twisted ladder or a double helix, DNA is composed of two long strands of nucleotides twisted around each other, each strand made up of a sugar-phosphate backbone and attached to one of four types of nitrogen bases: adenine (A), cytosine (C), guanine (G), and thymine (T).

The order of these bases forms the basis of our genetic code, with the sequence determining everything from the color of our eyes to our susceptibility to certain diseases. It's a language with only four letters, but these letters combine in myriad ways to form the sentences and paragraphs of our biological story.

Genes: The Units of Heredity

Within this vast expanse of DNA, specific sequences act as *genes*. A gene is a unit of heredity and is a region of DNA that encodes a functional RNA or protein product, and thus determines a particular characteristic in an organism. Genes are the instructions for building the proteins that carry out a variety of functions in a cell.

The human genome, which is the complete set of human DNA, includes about 20,000-25,000 genes. Although every cell in the human body contains a complete set of DNA, not all genes are active at all times. Gene expression, the process by which the instructions in our DNA are converted into a functional product, is a highly regulated process and is crucial for the survival and functioning of the organism.

DNA Replication and Inheritance

The process of DNA replication is central to the continuation of life. This meticulous process ensures that each new cell receives a complete and accurate copy of the genetic material. Errors in this process can lead to mutations, which can be benign, harmful, or occasionally beneficial, contributing to the process of evolution.

In sexual reproduction, DNA is inherited from two parents, with offspring receiving a mix of genes from both. This genetic combination is what makes each individual unique, barring identical twins.

Genetic Research and Its Implications

The study of DNA and genes has revolutionized our understanding of biology. The completion of the Human Genome Project in the early 21st century marked a significant milestone, opening the door for advancements in genomics, medicine, and biotechnology. Today, genetic research holds the promise of personalized medicine, where treatments and medications can be tailored to the individual's genetic makeup, revolutionizing healthcare and treatment strategies.

Ethical Considerations

As we delve deeper into genetic engineering and manipulation, ethical considerations come to the forefront. The potential to 'edit' genes through technologies like CRISPR brings both hope for curing genetic diseases and concerns about the implications of such power. The ethical, legal, and social implications of genetic research are areas of ongoing debate and are as important as the scientific discoveries themselves.

5.2 Genetic Engineering

Genetic engineering represents the pinnacle of human ingenuity in the field of biotechnology. It is the direct manipulation of an organism's genome using advanced methods and technologies. This chapter delves into the intricate world of genetic engineering, exploring its tools, applications, ethical implications, and the future prospects that it holds.

Advanced Tools in Genetic Engineering

CRISPR-Cas9 Technology: A paradigm shift in genetic manipulation, CRISPR-Cas9 allows for precise and efficient genome editing. The technology, derived from a natural defense mechanism in bacteria, has revolutionized our ability to edit DNA, offering unprecedented control over the genetic code.

Synthetic Biology: This interdisciplinary field combines engineering principles with biology. It involves the design and construction of new biological parts, devices, and systems, and the re-design of existing, natural biological systems for useful purposes. Synthetic biology is pushing the boundaries of genetic engineering, enabling the creation of entirely new organisms with tailor-made traits.

Cutting-edge Applications of Genetic Engineering

Genetic engineering stands at the forefront of scientific advancement, transforming numerous fields. Here, we delve into several of its most innovative and impactful applications.

Medical Therapeutics

Genetic engineering has revolutionized medicine, offering novel therapies and personalized treatment options.

Gene Therapy: Perhaps the most direct application in medicine, gene therapy involves correcting or replacing faulty genes. Recent advancements include the development of viral vectors that can safely and efficiently deliver therapeutic genes to target cells. An example is the use of adeno-associated viruses (AAVs) as vectors in treating genetic disorders like Spinal Muscular Atrophy (SMA).

Personalized Medicine: Genetic engineering enables treatments tailored to individuals' genetic makeup, particularly in oncology. Techniques like CRISPR are being used to modify patients' immune cells to target specific cancer types, as seen in CAR-T cell therapy for leukemia.

Regenerative Medicine: This field combines genetic engineering with tissue engineering to repair or replace damaged organs and tissues. Using induced pluripotent stem cells (iPSCs), scientists can reprogram adult cells into embryonic stem cell-like states, then coax them into different cell types needed for therapy.

Agricultural Enhancements

In agriculture, genetic engineering is used to develop crops with improved yield, nutritional quality, and resistance to pests and diseases.

Transgenic Crops: These are crops genetically modified to express a desirable trait, such as pest resistance or increased nutritional value. For example, Bt crops (like Bt corn) have been engineered to express a bacterial toxin that is harmful to specific pests, reducing the need for chemical pesticides.

Gene Editing in Agriculture: CRISPR and other gene-editing tools are revolutionizing agricultural biotechnology by enabling precise modifications without incorporating foreign DNA. This technology is used to improve crop resilience against climate change, enhance nutritional profiles, and increase yield.

Environmental BioSolutions

Genetic engineering provides innovative solutions to environmental problems, including pollution and conservation issues.

Bioremediation: This involves using genetically engineered microorganisms to degrade environmental pollutants. For instance, bacteria can be modified to break down oil spills or industrial waste, providing an efficient and eco-friendly cleanup method.

Bioenergy: Genetically modified organisms are also used to produce biofuels. Algae, for instance, can be engineered to produce higher lipid content for biodiesel production, offering a sustainable alternative to fossil fuels.

Conservation and Biodiversity: Genetic engineering offers tools for conservation biology, such as developing disease-resistant endangered species or using gene drives to control invasive species.

Industrial Applications

Beyond medicine and agriculture, genetic engineering

is making significant strides in various industries.

Biomanufacturing: The use of genetically modified microorganisms for the production of pharmaceuticals, enzymes, and other valuable chemicals is a growing field. This includes the production of insulin and other hormones using recombinant DNA technology.

Synthetic Biology: This emerging field involves creating new biological parts and systems or redesigning existing ones for useful purposes. Applications range from developing new biomaterials to constructing synthetic organisms that can perform specific tasks.

Neurogenetics and Brain Research

Genetic engineering is unlocking the mysteries of the brain and nervous system.

Gene Editing for Neurological Disorders: Techniques like CRISPR are being explored to treat genetic neurological disorders. This includes research into conditions like Huntington's disease and epilepsy.

Neuroprosthetics and Brain-Computer Interfaces (BCIs): Genetic engineering is contributing to the development of advanced neuroprosthetic devices that can interface with the brain, offering new hope for individuals with paralysis or brain injuries.

Ethical Considerations

Designer Babies:

The ability to edit genes raises profound ethical questions, particularly in the realm of human enhancement. The concept of "designer babies," where genetic traits such as intelligence, physical appearance, or even personality traits could be selected or modified, sparks intense ethical debates. As the technology advances, society grapples with defining the ethical boundaries of manipulating the human genome.

Environmental Concerns:

In agriculture, the widespread use of genetically modified crops has raised environmental concerns. The potential for unintended consequences, such as the development of resistant pests or the impact on non-target species, prompts a critical examination of the environmental implications of genetic engineering.

Regulatory Frameworks

Global Governance:

As genetic engineering advances, the need for robust regulatory frameworks becomes imperative. Nations and international bodies are challenged to establish guidelines that balance scientific progress with ethical and safety considerations. The absence of clear regulations raises concerns about the responsible use of genetic engineering technologies.

Public Engagement:

The ethical dimensions of genetic engineering necessitate active public engagement. Informed public discourse is crucial in shaping the policies and regulations that govern the application of genetic engineering. Striking a balance between scientific advancement and societal values requires collaboration between scientists, policymakers, and the general public.

Emerging Frontiers:

The future of genetic engineering holds exciting possibilities. From synthetic biology, where entirely new organisms are designed, to gene drives that alter entire populations of species, the boundaries of genetic manipulation continue to expand. As we step into this uncharted territory, questions of responsibility, transparency, and equity become ever more pressing.

A Call for Wisdom:

As the pages of genetic engineering's story unfold, there is a collective call for wisdom in wielding this unprecedented power. Balancing scientific curiosity with ethical considerations, ensuring equitable access to the benefits of genetic engineering, and fostering responsible innovation are essential for navigating the intricate landscapes of our genetic future.

In conclusion, genetic engineering represents a double-edged sword, offering immense potential for human betterment while demanding a cautious approach to navigate the ethical and societal implications. As we stand at the crossroads of discovery, the choices we make today will shape the course of humanity's genetic journey for generations to come.

Chapter 6: The Story of Life

6.1: The Origin of Life

Prelude to Existence; The cosmic drama that unfolded eons ago laid the foundation for the spectacular phenomenon we call life.

In the vast expanses of the universe, the cosmic stage witnessed the fusion of elemental ingredients that would ultimately give rise to the intricate dance of living organisms on Earth.

Cosmic Ingredients

The story begins with the elemental ballet orchestrated in the hearts of stars. Hydrogen and helium, the cosmic primordials, fused in the intense heat and pressure within stellar cores, giving birth to heavier elements. As these stars reached the end of their life cycles, they unleashed these elements into space through dazzling explosions—supernovae. This stardust, rich in carbon, oxygen, nitrogen, and other essential elements, became the cosmic building blocks of life.

These elements, dispersed across the cosmos, eventually found their way to a young planet orbiting an average star in an unremarkable corner of the Milky Way galaxy—Earth.

Building Blocks of Life

Life, as we recognize it, is fundamentally based on carbon compounds. Carbon's unique ability to form stable bonds with a variety of other elements and itself allows for the creation of diverse and complex molecules. This characteristic made carbon the linchpin for the organic chemistry that underlies the biological processes of all living organisms.

The dance of these carbon molecules on the early Earth set the stage for the alchemy of life to unfold.

Early Earth: A Crucible of Possibilities

Earth, approximately 4 billion years ago, was a radically different planet from the one we know today. The surface was dominated by vast oceans of primordial soup—a concoction rich in organic molecules, amino acids, and other prebiotic compounds. The atmosphere, devoid of the oxygen we breathe today, consisted of methane, ammonia, water vapor, and other gases.

Primeval Oceans and Atmosphere

In this primordial environment, the stage was set for the chemistry of life to emerge. The oceans became a vast laboratory where the elements of life could mingle and react, fostering the creation of increasingly complex molecules.

The Miller-Urey Experiment

The groundbreaking experiment conducted by Stanley Miller and Harold Urey in the 1950s provided a glimpse into the conditions of early Earth. They simulated the atmosphere of the time, subjected it to electrical discharge mimicking lightning, and observed the formation of amino acids—the building blocks of proteins. This experiment not only demonstrated the plausibility of life's spontaneous emergence but also ignited a new era of scientific inquiry into the origins of life.

The Enigma of Life's First Steps

RNA World Hypothesis

As scientists strive to unravel the mystery of life's origins, the RNA World hypothesis emerges as a compelling narrative. This theory posits that before the emergence of DNA, RNA played a central role. RNA, with its ability to store genetic information and catalyze chemical reactions, could have been the precursor to the more stable DNA.

The transition from simple organic molecules to self-replicating entities represents a critical juncture in the enigma of life's first steps. How did life move from the haphazard dance of molecules to the elegant choreography of self-replicating structures?

Protocells and Membranes

Life requires boundaries, a protective membrane that separates internal processes from the external environment. The formation of protocells—a term coined for simple, self-organizing structures with a rudimentary membrane—marks a pivotal step in life's origin story. The challenge lies in understanding how these early protocells gained the ability to replicate and evolve.

Evolutionary Pioneers

First Microbes

Around 3.5 billion years ago, the curtain rose on the first act of the evolutionary play. Simple, single-celled organisms, the pioneers of life, made their debut. These microbes thrived in the ancient seas, laying the foundation for the remarkable diversity and complexity that would characterize life on Earth.

Evolutionary Milestones

The journey of life is marked by significant milestones, each contributing to the rich tapestry of biological evolution. The development of photosynthesis, a process that converts sunlight into energy and produces oxygen as a byproduct, transformed Earth's atmosphere and paved the way for oxygen-breathing organisms.

The emergence of multicellular life forms added another layer of complexity, setting the stage for the myriad of life forms that would follow—from microscopic organisms to the towering giants that roam the Earth today.

The Unfinished Tapestry

Continuing the Story

The story of life's origin is far from complete; rather, it is an ongoing narrative that extends across geological epochs. Scientific inquiry, fueled by technological advancements and interdisciplinary collaboration, continues to peel back the layers of time, revealing new chapters and rewriting others.

Astrobiology and Extraterrestrial Life

The search for life extends beyond Earth. Astrobiology, a field at the intersection of biology, chemistry, and astronomy, explores the conditions that could support life on other planets and moons in our solar system and beyond. The discovery of extremophiles—

organisms thriving in extreme environments on Earth—fuels the imagination with possibilities of life in seemingly inhospitable corners of the cosmos.

Philosophical Reflections

Beyond the scientific intricacies, the quest to understand life's origin invites philosophical reflections. What is the nature of life? What is its purpose? The exploration of life's origins transcends the empirical realm, inviting contemplation and awe.

As we delve into the depths of our biological heritage, we are confronted with the profound mystery of existence. The origin of life, a cosmic symphony played out on the stage of our planet, beckons us to contemplate our place in the vast tapestry of the universe.

In exploring the origin of life, we embark on a journey through time and space, tracing the footsteps of our ancient ancestors. The pages of this chapter unfold the early chapters of life's grand narrative, inviting us to marvel at the cosmic forces that birthed existence and ponder the profound mystery of life's emergence on our planet.

Deepening the Narrative

Emergence of Complexity

As the curtain rose on the evolutionary stage, life embraced the challenge of complexity. From the simple structures of early microorganisms, life gradually evolved into more intricate forms. The mechanisms of natural selection and adaptation sculpted life's journey, giving rise to diverse species, each finely tuned to its environment.

Cambrian Explosion

Approximately 541 million years ago, the Earth witnessed the Cambrian Explosion—a geological heartbeat that saw the rapid diversification of multicellular life. This event marked a pivotal moment in evolutionary history, introducing an array of body plans and biological innovations that set the stage for the astonishing biodiversity we see today.

Coevolution and Interconnectedness

Life's story is not merely a sequence of isolated events but a tapestry woven through coevolution and interconnectedness. Species, in their struggle for survival and reproduction, influence each other's evolutionary trajectories. From the delicate dance between flowers and pollinators to the predator-prey relationships shaping ecosystems, life weaves an intricate web of dependencies.

Symbiosis and Mutualism

Symbiotic relationships, where different species live in close association, highlight the collaborative nature of evolution. Examples abound, from the mitochondria within our cells—believed to have originated from ancient symbiotic bacteria—to the mutually beneficial partnerships between plants and mycorrhizal fungi. Life, it seems, thrives on collaboration as much as competition.

Challenges and Extinctions

The Ebb and Flow of Life

Life's journey is not without its challenges. The planet has witnessed multiple mass extinctions, each a reset button that paved the way for new forms of life to emerge. The most infamous of these, the Cretaceous-Paleogene extinction event around 66 million years ago, spelled the end for dinosaurs and opened opportunities for mammals to dominate terrestrial ecosystems.

Adaptation and Resilience

In the face of challenges, life demonstrates remarkable adaptive capabilities. Extant species carry within them the genetic imprints of survival through tumultuous times. The ability to adapt to changing environments, coupled with the relentless drive to reproduce, has ensured the continued existence of life in its myriad forms.

Human Emergence: A Pinnacle or Challenge?

The story of life takes an intriguing turn with the

emergence of Homo sapiens—the species capable of reflecting on its own origin. Human evolution is marked by cognitive prowess, toolmaking, and the ability to manipulate the environment. As humanity forges its path, questions arise about our role as stewards of the planet and the impacts of our actions on the web of life.

The Story Continues

Beyond Earth: Astrobiology and Beyond

The search for life extends beyond our home planet. As probes explore the moons of Jupiter and Saturn and telescopes peer into the atmospheres of exoplanets, astrobiology aims to uncover the conditions conducive to life elsewhere in the cosmos. The prospect of finding extraterrestrial life, even in microbial form, tantalizes the human imagination and fuels our exploration of the universe.

Technological Evolution

In the tapestry of life's narrative, the emergence of intelligent, technologically advanced beings raises questions about the trajectory of evolution. From the controlled use of fire to the exploration of outer space, humanity's journey is marked by technological innovation. As we look to the future, the prospect of artificial intelligence and biotechnological advancements adds new layers to the evolving narrative of life.

A Tapestry of Wonder

The Poetry of Existence

The story of life is more than a scientific narrative; it is a poetry of existence. It is a tale of resilience, adaptation, and the ceaseless quest for survival. Life's tapestry, woven across epochs, reflects the grandeur of the cosmos and the intricate dance of matter and energy.

Ethical Considerations

As stewards of life on Earth, humans are faced with ethical considerations. The power to manipulate the genome, reshape ecosystems, and explore the cosmos carries profound responsibilities. How we navigate this intersection of knowledge and power will shape the future chapters of the story of life.

Conclusion

In exploring the origin of life, we uncover not just scientific facts but profound insights into our own existence. From the cosmic cauldron that birthed our elements to the intricate dance of coevolution, life's story is a testament to the beauty and complexity of the universe.

As we stand on the threshold of the unknown, the unfolding narrative of life invites us to marvel at the wonders of the cosmos and reflect on our place in this grand tapestry. The story continues, and with each discovery, each ethical choice, humanity becomes an integral part of the ongoing epic—the ever-unfolding story of life.

6.2: Early Life Forms

Prologue to Primordial Life

The emergence of life on Earth is an ancient narrative, a tale woven through the fabric of geological epochs. As we delve into the enigmatic epochs that precede the rise of complex organisms, we embark on a journey to explore the dawn of life—the era when the first stirrings of existence unfolded in the vast cradle of Earth's early oceans.

Geological Epochs and Life's Incunabula

The Precambrian eon, spanning roughly 4 billion years of Earth's history, encompasses the infancy of life. This vast expanse, often referred to as life's incubator, witnessed the tentative steps of microbial life forms, paving the way for the intricate tapestry of biodiversity that would later grace the planet.

Microbial Odyssey: The Prokaryotic Epoch

The microbial odyssey begins with the emergence of prokaryotic life forms—organisms lacking a distinct nucleus and membrane-bound organelles. Archaebacteria and eubacteria, the precursors to modern archaea and bacteria, embarked on a journey of resilience and adaptation in Earth's ancient seas.

Prokaryotic Prowess: Extremophiles and Anaerobes

Prokaryotes demonstrated an astonishing resilience to extreme environments. Extremophiles, thriving in environments deemed inhospitable by conventional standards, showcased the adaptive prowess of early life. Anaerobic bacteria, devoid of the need for oxygen, flourished in an atmosphere that lacked this vital element, laying the foundation for the subsequent evolution of oxygenic photosynthesis.

Oxygen Revolution: The Cyanobacterial Triumph

The Phototropic Symphony

Approximately 2.4 billion years ago, a pivotal chapter in Earth's biological narrative unfolded—the advent of cyanobacteria. These ingenious microorganisms harnessed the power of sunlight through photosynthesis, fundamentally altering Earth's atmospheric composition. The Oxygen Revolution marked a transformative epoch, paving the way for the evolution of aerobic respiration and setting the stage for the rise of complex life forms.

Oxygen as a Double-Edged Sword

While oxygen became the breath of life for aerobic organisms, its accumulation posed challenges for anaerobic life forms. The

Great Oxygenation Event, a consequence of cyanobacterial photosynthesis, triggered mass extinctions among anaerobes, illustrating the complex interplay between the evolution of life and the changing conditions of the biosphere.

Eukaryotic Emergence: The Complexity Unveiled

The emergence of eukaryotic cells, distinguished by a defined nucleus and membrane-bound organelles, remains one of the most profound transitions in the history of life. Eukaryogenesis, the process by which eukaryotic cells evolved from prokaryotic precursors, is akin to a molecular ballet—an intricate dance of genetic material and cellular structures.

Endosymbiotic Symphony

The endosymbiotic theory, proposed by Lynn Margulis, provides a captivating narrative for the origin of eukaryotic organelles. Mitochondria, the cellular powerhouses, and chloroplasts, the sites of photosynthesis, are believed to have originated through symbiotic relationships with ancestral prokaryotes. This mutualistic dance of cellular entities laid the foundation for the rise of complex, multicellular life.

The Ediacaran Enigma

The Ediacaran period, approximately 635 to 541 million years ago, is a canvas shrouded in mystery. Fossilized imprints of soft-bodied organisms, etched into the sedimentary layers, unveil a glimpse of life forms that preceded the explosion of complex multicellular organisms during the Cambrian period.

Iconic Fauna: From Imprints to Interpretations

The Ediacaran biota, characterized by enigmatic forms such as Dickinsonia and Spriggina, defies easy classification. The fossil record raises questions about the ecological roles, lifestyles, and evolutionary relationships of these ancient organisms. The Ediacaran enigma continues to captivate paleontologists and sparks debates about the roots of complex animal life.

The Cambrian Kaleidoscope

Radiant Diversity: Explosion of Body Plans

The Cambrian Explosion, spanning a mere 20-25 million years, heralded a burst of biological creativity. The fossil record from this period reveals a kaleidoscope of diverse body plans, marking the rapid evolution of complex multicellular organisms. Arthropods, mollusks, and chordates emerged as representatives of the early animal kingdom.

Evolutionary Arms Race: The Drive for Adaptation

The Cambrian Explosion was not just a display of diversity but an evolutionary arms race. Intense competition for resources, ecological niches, and strategies for predation and defense spurred the development of novel adaptations. The quest for survival drove the refinement of sensory organs, the evolution of exoskeletons, and the emergence of complex behaviors.

Reconstructing Ancient Biomes

Paleogeographic Puzzles

Reconstructing ancient biomes involves navigating through paleogeographic puzzles. The positioning of continents, the configurations of oceans, and the climate of bygone eras all play pivotal roles in shaping the habitats that influenced the evolution of early life forms.

Ediacaran Landscapes and Cambrian Ecologies

The landscapes of the Ediacaran and Cambrian periods set the stage for the drama of evolution. Ediacaran seafloors hosted soft-bodied organisms, while the Cambrian seas witnessed the rise of more complex, mobile fauna. The interplay between environmental factors and evolving life forms is a complex tapestry awaiting decipherment.

The End of the Chapter: A Glimpse of the Future

Mass Extinctions and Evolutionary Cascades

As this chapter in the saga of life draws to a close, it is impossible to ignore the specter of mass extinctions. These cataclysmic events, from the Ordovician-Silurian Extinction to the Permian-Triassic Extinction, shaped the course of evolution. Yet, within the shadows of devastation, new opportunities for evolutionary innovation arose, leading to the eventual emergence of new life forms.

Evolutionary Trajectories: A Tapestry Unraveling

The story of early life forms is not a linear narrative but a tapestry of branching trajectories.

Evolutionary experiments, some successful and others short-lived, have left an indelible mark on the tree of life. As we glimpse the pages of ancient history, we anticipate the untold chapters that lie ahead—an unfolding narrative shaped by the ever-evolving dance of life.

In the intricate dance of evolution, early life forms set the stage for the grand theater of biodiversity that spans geological epochs. From the humble beginnings of prokaryotic microorganisms to the dazzling display of body plans during the Cambrian Explosion, the narrative of early life forms unveils the profound complexities that underpin the biological heritage of our planet. As we navigate through the fossilized imprints and molecular mysteries, we gain a deeper appreciation for the resilience, adaptability, and sheer creativity inherent in the journey of life.

The next chapter awaits, promising further revelations and a continued exploration of life's unfolding tapestry.

6.3: Evolutionary Milestones

As life's journey unfolds, the symphony of adaptation plays a central role in shaping the diversity of organisms. Natural selection, the guiding force of evolution, orchestrates the march of adaptation. The interplay between genetic variation, environmental challenges, and reproductive success composes a harmonious melody that echoes across generations.

Adaptive Radiations: Nature's Exuberance

Evolutionary history is punctuated by moments of exuberant creativity known as adaptive radiations. These periods witness a rapid diversification of species into a variety of ecological niches. From the explosive radiation of mammals after the extinction of dinosaurs to the proliferation of finches on the Galápagos Islands, adaptive radiations showcase nature's ability to innovate in response to changing landscapes.

The Symphony of Coevolution

Mutualistic Duets

Coevolution, the reciprocal influence between different species, adds intricate melodies to the evolutionary symphony. Mutualistic relationships, where two species evolve in a way that benefits both, exemplify the collaborative nature of coevolution. Examples abound, from the intimate partnership between flowering plants and pollinators to the synergistic dance of cleaner fish and their clients on coral reefs.

Evolutionary Arms Races: Predator and Prey

In the arms race between predator and prey, the evolutionary symphony crescendos with adaptations and counter-adaptations. Camouflage, mimicry, and defensive mechanisms evolve in a perpetual struggle for survival. The evolutionary arms race is a dynamic dialogue that shapes the anatomical and behavioral repertoire of organisms, revealing the intricate strategies of the hunt and the art of evasion.

The Rise of Tetrapods: Conquering Land

The Amphibian Interlude

The transition from water to land marks a pivotal chapter in evolutionary history. Amphibians, with their dual life cycles adapted to both aquatic and terrestrial environments, pioneered the conquest of land. This interlude set the stage for the evolution of tetrapods—four-limbed vertebrates—ushering in a new era of ecological possibilities.

Tetrapod Triumph: From Fins to Limbs

The emergence of tetrapods witnessed the transformation of fins into limbs—an anatomical innovation that opened novel avenues for exploration and survival. The Devonian period, often dubbed the Age of Fishes, saw the rise of creatures like Tiktaalik, a transitional form that bridged the gap between fish and tetrapods. The conquest of land brought forth a symphony of evolutionary experimentation and adaptation.

Wings of Innovation: The Avian Ascent

The Dawn of Feathers

The evolution of feathers, initially adapted for insulation and display, set the stage for one of nature's most extraordinary innovations—the development of flight. Avian evolution, characterized by the rise of birds, showcases the transformative power of natural selection. The feathers that once adorned dinosaurs became the wings that allowed birds to soar through the skies.

Adaptive Radiations in the Aerial Realm

The sky became a canvas for evolutionary artistry as birds radiated into diverse ecological niches. From the acrobatic maneuvers of hummingbirds to the powerful soaring of eagles, the aerial realm witnessed a symphony of adaptations. Flight not only provided access to new habitats but also fueled the evolution of specialized beaks, talons, and behaviors.

Mammalian Marvels: From Pouches to Brains

The Marsupial Mastery

Mammals, distinguished by their ability to nurse their young with milk, exhibit a remarkable diversity of reproductive strategies. The marsupial lineage, exemplified by creatures like kangaroos and opossums, embraces a unique reproductive method. Marsupials, born in an embryonic state, complete their development in pouches—an adaptation that showcases the diversity of maternal care strategies.

Placental Pinnacle: A Symphony of Gestation

The evolution of placental mammals represents a pinnacle of reproductive innovation. The development of a placenta enables these mammals to nourish their offspring within the womb for an extended period. This adaptation fosters intricate social structures, complex behaviors, and the evolution of large-brained species. From the echolocation prowess of bats to the cognitive complexity of primates, placental mammals compose a symphony of evolutionary marvels.

Human Odyssey: An Evolutionary Opus

The evolution of Homo sapiens is a captivating opus in the grand symphony of life. Bipedalism, the ability to walk on two legs, is a hallmark adaptation that distinguishes humans from their primate relatives. The fossil record reveals the gradual transition from quadrupedal ancestors to the upright posture of early hominins—an evolutionary ballet that opened new horizons for exploration and tool use.

Cognitive Crescendo: The Human Mind

The ascent of humans is marked by a cognitive crescendo—a symphony of neurological complexity and cultural innovation. The enlargement of the human brain, coupled with the development of language and symbolic thought, propelled Homo sapiens to unprecedented heights of intellectual prowess. The evolution of the human mind is a testament to the power of adaptive intelligence and the cultural transmission of knowledge.

The Genomic Sonata: From DNA to CRISPR

Genomic Orchestration

The unraveling of the genomic symphony has revealed the molecular melodies that underpin the diversity of life. The genomic orchestra, conducted by the intricate dance of DNA, orchestrates the development and functioning of organisms. From the genetic code's orchestration of proteins to the regulatory dynamics that shape traits, genomics unveils the fundamental mechanisms of evolution.

CRISPR: Editing the Genetic Score

In the 21st century, the genetic symphony encounters a revolutionary crescendo with the advent of CRISPR-Cas9 technology. This molecular tool, inspired by the microbial immune system, allows for precise editing of the genetic score. The ability to modify genes opens new frontiers in genetic engineering, challenging ethical boundaries and reshaping the future of evolution.

Epilogue: Harmony in Flux

As we reflect on the evolutionary milestones that have shaped life's journey, we recognize that the symphony of adaptation and innovation is an ever-evolving composition. The harmony of life is in flux, with each species contributing its unique notes to the grand orchestration. From the microbial realms to the heights of human cognition, the symphony continues to resonate across the epochs.

Ethical Overtures

The exploration of evolution and its milestones invites contemplation on the ethical dimensions of our role in shaping the evolutionary narrative. As stewards of the planet, humans bear a responsibility to navigate the complexities of genetic engineering, environmental conservation, and the preservation of biodiversity. The ethical overtures of our actions resonate in the harmonies of the evolving symphony, influencing the trajectory of life's ongoing opus.

In this exploration of evolutionary milestones, we traverse the landscapes of adaptation, coevolution, and transformative innovations. From the ancient rhythms of natural selection to the modern crescendo of genetic editing, the symphony of evolution continues to captivate and inspire. As we stand at the intersection of biological past and future, the harmonies of life echo through the corridors of time, inviting us to appreciate the intricate melodies that have shaped the vast tapestry of biodiversity on our planet.

Conclusion

In our exploration of the mysteries that shape our existence, from the vast reaches of the cosmos to the intricate tapestry of life on Earth, we've embarked on a journey of discovery. The chapters have unfolded like chapters of a cosmic novel, revealing the wonders and complexities that define our reality. As we reach the final notes of this symphony of exploration, it's fitting to pause, reflect, and consider the profound implications of our understanding.

The Universe, with its galaxies, black holes, and dark matter, has whispered its secrets through the cosmic winds. We've marveled at the birth and fate of stars, realizing that in the grand dance of the cosmos, we are but stardust, intricately connected to the very fabric of the universe. The Solar System, our celestial neighborhood, has unfolded its planetary mysteries, inviting us to appreciate the unique characteristics of each planetary neighbor and ponder the enigmas that still linger in the cosmic expanse.

Descending to our planetary home, we've delved into the Earth and its systems, navigating the realms of geology, plate tectonics, and the dynamic biosphere. The Earth, with its ever-changing climate and water cycles, has proven to be a stage for the drama of life. It is on this stage that the intricate dance of evolution and biodiversity unfolds.

The story of life on Earth, from the ancient rhythms of natural selection to the modern crescendo of genetic editing, has been nothing short of a captivating symphony. We've traced the origin of life, marveled at early life forms, and witnessed evolutionary milestones that have shaped the biodiversity we cherish today. But as the final chords of this evolutionary opus resonate, we are compelled to contemplate the ethical dimensions of our role in shaping the narrative.

The ethical overtures of our actions reverberate through the corridors of time, influencing the trajectory of life's ongoing opus. As stewards of the planet, we bear a profound responsibility to navigate the complexities of genetic engineering, environmental conservation, and the preservation of biodiversity. The harmonies of evolution, from adaptation to coevolution and transformative innovations, echo through the landscapes we traverse. Our choices, like musical notes, contribute to the ever-evolving symphony, leaving an indelible mark on the melody of life.

As we stand at the intersection of the biological past and future, the harmonies of life echo through the corridors of time, inviting us to appreciate the intricate melodies that have shaped the vast tapestry of biodiversity on our planet. It's a call to action, a call to mindfulness in our endeavors as caretakers of Earth. The interconnectedness of all life forms is a melody that transcends species, borders, and time itself.

In this closing chapter, we stand not at an endpoint but a

crossroads—the knowledge gained now a guiding light for future exploration. The universe unfolds its mysteries, and the story of life on Earth remains an ever-unfolding narrative. Our understanding, a torch for curious minds and eager hearts to follow in exploration's footsteps

As we close the pages of "How The World Really Works," let us carry forward the spirit of curiosity, the reverence for the unknown, and the responsibility to tread lightly on the delicate balance of our planet. The chapters of this book are not mere words on paper; they are waypoints in the grand adventure of understanding our existence.

So, to the readers who have joined this odyssey, I extend my deepest gratitude. May the mysteries of the universe continue to captivate your imagination, may the wonders of our planet inspire awe, and may the ethical overtures embedded in the symphony of evolution guide your path. As we conclude this chapter, let us embrace the unknown with open minds, for it is in the pursuit of knowledge that we contribute to the ever-evolving melody of the cosmos.

"Life's dance, a sonnet of understanding entwined with the ballet of stewardship."

Manufactured by Amazon.ca
Acheson, AB